D1259262

Scientific Computation

Springer
Berlin
Heidelberg
New York
Barcelona
Hong Kong
London
Milan
Paris
Singapore
Tokyo

Physics and Astronomy

ONLINE LIBRARY

http://www.springer.de/phys/

Scientific Computation

Series homepage – http://www.springer.de/phys/books/sc/

Aristide Mbiock   Roman Weber

Radiation in Enclosures

Elliptic Boundary Value Problem

With 55 Figures and 20 Tables

 Springer

Dr. Aristide Mbiock
Dept. of Applied Physics
Thermal and Fluids Sciences
Delft University of Technology
Lorentzweg 1
2628 CJ Delft, The Netherlands

Dr. Roman Weber
Research Station b. v.
International Flame Research Foundation
Wenckebachstraat 1
1951 JZ Velsen-Noord, The Netherlands

Library of Congress Cataloging-in-Publication Data

Mbiock, Aristide, 1969-
    Radiation in enclosures : elliptic boundary value problem /
Aristide Mbiock, Roman Weber.
        p.   cm. -- (Scientific computation, ISSN 1434-8322)
    Includes bibliographical references and index.
    ISBN 3-540-66095-X (hardcover : alk. paper)
    1. Heat--Transmission--Mathematical models. 2. Boundary value
problems--Numerical solutions. 3. Differential equations, Elliptic-
-Numerical solutions.   I. Weber, Roman, 1951-   . II. Title.
III. Series.
QC320.M394 1999
53C'.2'0015118--dc21   99-34218
                                                               CIP

ISSN 1434-8322
ISBN 3-540-66095-X Springer-Verlag Berlin Heidelberg New York

Springer-Verlag is a company in the specialist publishing group BertelsmannSpringer
© Springer-Verlag Berlin Heidelberg 2000
Printed in Germany

The use of general descriptive names, registered names, trademarks, etc. in this publication does not imply, even in the absence of a specific statement, that such names are exempt from the relevant protective laws and regulations and therefore free for general use.

Typesetting: Camera-ready copy from the authors
Cover design: *design & production* GmbH, Heidelberg
Printed on acid-free paper   SPIN: 10717227   55/3144/mf - 5 4 3 2 1 0

We dedicate this book to

Michael, Ritha, Josine, Leonie and Marie

Preface

During the last half century, the development and testing of prediction models of combustion chamber performance have been an ongoing task at the International Flame Research Foundation (IFRF) in IJmuiden in the Netherlands and at many other research organizations. This task has brought forth a hierarchy of more or less standard numerical models for heat transfer predictions, in particular for the prediction of radiative heat transfer. Unfortunately all the methods developed, which certainly have a good physical foundation, are based on a large number of extreme simplifications or uncontrolled assumptions. To date, the ever more stringent requirements for efficient production and use of energy and heat from combustion chambers call for prediction algorithms of higher accuracy and more detailed radiative heat transfer calculations. The driving forces behind this are advanced technology requirements, the costs of large-scale experimental work, and the limitation of physical modeling. This interest is growing more acute and has increased the need for the publication of a textbook for more accurate treatment of radiative transfer in enclosures.

The writing of a textbook on radiative heat transfer, however, in addition to working regularly on other subjects is a rather difficult task for which some years of meditation are necessary. The book must satisfy two requirements which are not easily reconciled. From the mathematical point of view, it must be written in accordance with standards of mathematical rigor and precision. From the radiative heat transfer point of view, it must not be written so abstractly as to discourage physicists and engineers who need this mathematical tool. We have consequently made an effort to prepare a coherent and comprehensive story to meet the need of advanced heat transfer work. We aim to make the subject easier to grasp, not only by giving proofs, but also by showing the interconnections with other fields of physics and mathematics, and actual numerical calculations in heat transfer.

This book describes the latest IFRF developments [1] and is concerned with the formulation and solution of the radiative heat transfer problem in enclosures. The discussion here is firmly directed towards the relevant basis of a mathematically exact method for computing radiation heat transfer in enclosures as opposed to a merely formal improvement of existing methods. The textbook is essentially self-contained and is intended for anyone who, having passed through a basic course of heat transfer, wishes to study the special case of radiation. Thus, we have attempted to produce a balanced work that is both useful to the practising design engineers and stimulating to researchers in industry and universities.

The book consists of nine relatively detailed chapters so that they may serve as source for some of the more subtle points of radiation theory. A global view of the scope covered is given in the table of contents. Here, we merely point out some particulars.

We lay the foundations in the first chapter by recalling the concept of radiant energy and by surveying the major events in the development of methods for radiative heat transfer calculations. An outline of the aim and the coverage of topics follow this historical survey. We also recall some concepts of boundary value problems and we underline the basis for its variational solution. The methodology presented in this chapter is then implemented and detailed in the remaining chapters.

The second chapter describes the physical model considered, and the core of the underlying situation. In this chapter, the equation of transfer for local intensities and its integral along a single line of sight are derived. The properties pertinent to radiation and the radiative flux are examined. The energy balance relations are then presented and the equation of transfer is formulated for local radiative heat fluxes.

The third chapter reviews the developments in the computational methods presently used in engineering radiative heat transfer applications. These methods are roughly classified into those based on the integrated directional equation of transfer and those based on the net energy balance relations. This review is followed by a description of the equation studied further on.

The fourth chapter focuses on establishing a unified theory for the solution of the radiative transfer equation in enclosures. In this chapter, a formulation of a satisfactory mathematical problem and a study of the analytical properties of the solution are developed. Also, a considerable part of this chapter is devoted to the questions of existence, uniqueness and analytical form of the theoretical solution.

The fifth chapter is concerned mainly with a representation of the solution in terms of data and related questions, and presents detailed analysis on the numerical solution. In this chapter, an approximate problem of radiant energy transfer is formulated. Existence and uniqueness of the associated solution are demonstrated on the basis of variational principle and finite element method, and the algorithm for numerical calculations is outlined.

The sixth chapter resumes some specific cases of radiative transfer. This brief résumé is followed in the seventh chapter by a description of the theoretical models often used for computing the spectral properties of infrared-radiating gas species encountered in industrial furnaces.

The eighth chapter, on an experimental furnace, illustrates the application of the solution procedure for calculating fluxes to the heat sink. To this end, the radiant energy exchanges are computed for four flames: three natural gas flames and a propane flame. The flames were generated in the Furnace No. 1 of the IFRF. The chapter provides a comprehensive validation of the mathematical procedure developed against the measured data (fluxes).

The book is concluded in chapter nine with both a discussion of radiative heat transfer in systems where the effects of scattering onto the overall radiation exchange is not negligible and the coupling of radiation with conduction and/or convection heat transfer.

For consistency and clarity, we use a functional notation to list explicitly the variables upon which a given quantity depends. We have tried as far as possible to give a simple and self-contained treatment of the problem of heat transfer by radiation in enclosures. We are apologetically conscious and understand that much remains to be done in the field of radiation, particularly on the medium properties related to the absorption coefficient. Still, there is no doubt that this book in its present form will be an efficient tool for advanced students, scientists and engineers with interest in radiative heat transfer. The book as presented now is certainly uneven in style and completeness. Yet, we hope that it will be a stimulus to further developments in the field of advanced heat transfer in enclosures.

The completion of this book, which was undertaken at the International Flame Research Foundation in IJmuiden in the Netherlands, would have been impossible without the sustained assistance given to us by a number of people. We are indebted to both Peter Roberts and Willem van de Kamp for the support we received throughout a period of almost four years of work on radiation heat transfer.

We gratefully acknowledge the computer facilities and documentation provided by the IFRF, the financial support, and the permission from the Joint Committee to use the Foundation's data and to publish this book. We have likewise to thank Springer-Verlag, particularly Prof. Dr.-Ing. Wolf Beiglöck, who have undertaken the task of publication.

We are pleased to express our gratitude to Prof. Ryszard Bialecki, from the Institute of Thermal Technology at Silesian Technical University in Poland, for productive suggestions and comments on the engineering and numerical aspects of radiation transfer. Actually in 1989, It was Prof. Ryszard Bialecki and Prof. Andrzej Nowak who drew our attention to the boundary element method and its potential for radiation exchange calculations.

We thank Prof. André Draux, from the National Institute of Applied Sciences in Rouen in France, for valuable clarifications on the mathematical aspects of heat transfer equations.

Our special thanks are due to Prof. Dr.-Ing Wolfgang Wendland, for his patience and understanding when confronted with receiving us at the Mathematisches Institut of Stuttgart University in Germany. To him we owe much more than can be expressed. He helped us enormously by performing an excellent and attentive review on the mathematical theory presented in this book.

We are also very much obliged to all the IFRF investigators and the technical team with whom we had direct contact whilst working at the Research Station.

To all these persons and to those whose names may have been omitted, we extend our profound thanks.

IJmuiden, August 1999   Aristide Mbiock & Roman Weber

Contents

1
Introduction

All materials viewed at the microscopic scale consist of atoms or molecules packed closely together as in a solid or a liquid, or sparsely distributed as in a gas. From one region of a material to another region, heat can be transferred by one of the three processes: convection, conduction, or radiation. Convection is the carrying of heat by movement of a fluid; in convection, the energy transfer results from movement of heated or energy containing matter from one point to another. The physical mechanism of conduction is quite complex. Its includes such varied phenomena as molecular collisions in gases, lattice vibrations in crystals, and movement of free electrons in metals. The last process, radiation, is the subject of this book. In contrast to convective and conductive heat transfer, for radiant exchange to occur no medium needs to be present. The radiant energy passes through a perfect vacuum, hence its basic laws are independent of the properties of any particular substance. When no medium is present, radiation is the only mode of energy transfer.

1.1 Thermal Radiation

Radiation occurs when a molecule of matter emits or absorbs a train of electromagnetic waves, or photons, by lowering or raising its molecular energy levels. In equilibrium state, the energy and wavelengths of emission depend on the temperature of the material. Figure 1.1 delineates the spectrum of electromagnetic radiation.

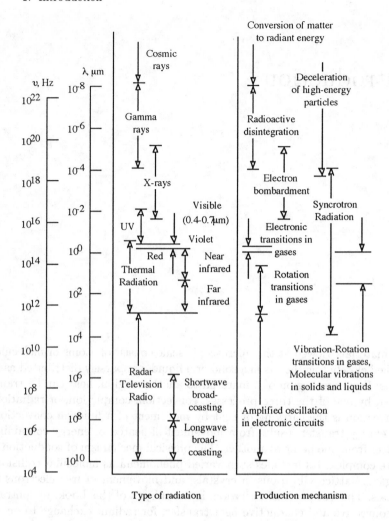

Figure 1.1. Spectrum of electromagnetic radiation

As the temperature of the material increases, the wavelength of the predominant electromagnetic phenomena decreases and we move from the infrared region to the visible region, and on to the ultraviolet region. As the wavelength decreases with increasing temperature, the rate of emission of electromagnetic energy increases and heat transfer by radiation becomes more significant.

It is the intermediate range of the spectrum, where radiation is detected as heat or light, which extends from approximately 0.1 to 100 μm and includes a portion of the ultraviolet and all the visible and infrared, that is pertinent to heat transfer by radiation in enclosures. This wavelength region gives mankind heat, light, photosynthesis, and all their attendant benefits. Everyday examples of such radiation heat transfer include the heating effect of sunshine on a clear day, or the effect of a fire on the human body's side facing it. Another effect during a cold winter night is that we feel more comfortable in a room whose curtains are drawn than in a room heated to the same temperature and with open curtains.

The short wavelength gamma rays, X rays, and ultraviolet radiation are of interest to high-energy and nuclear process, while the long wavelength microwaves and radio waves are of concern to electrical processes.

Radiation heat transfer is of importance in energy production and conversion, as well as in the process industry, steel making sector and chemical industry. One factor that accounts for this importance is the manner in which radiant energy emission depends on the temperature. In most applications, convective and conductive energies are almost linearly proportional to the temperature difference to the first power whereas radiant energy is proportional to the differences in temperature to the fourth power. Thus, with rising the temperature levels, the radiant energy transfer becomes important and may be totally dominant over convection and conduction. Some examples include combustion applications (e.g. boilers, furnaces, turbines and diesel engines, fluidized bed combustors, etc.) or nuclear reactors. In these applications, there is an incentive to operate processes at very high temperatures where the thermodynamic efficiency is improved and the chemical reactions are carried more closely to completion.

1.2 Short Historical Background

Over the last fifty years, the trends toward increased unit size and heat loading of industrial furnaces have emphasized the need for methods of prediction of furnace performance and heat transfer. Heat transfer, including radiation, became a subject taught in engineering schools and universities. Nowadays many industrial design problems for energy and heat production require consideration on how to either reduce or enhance it, in order to achieve high thermal efficiency.

Radiation is the dominant mode of energy transport in most industrial furnaces and combustion chambers. Here, the heat transmitted by radiation amounts to 60-90% of the overall energy transfer depending on the design. Until the early 1920s radiation was treated as a correction to the convective heat transfer. Its importance in industry was recognized in the 1920s as a mechanism of heat transfer in furnaces [2]. In the 1920s, the need for design tools that could adequately predict radiant energy transfer inside furnaces gradually became apparent.

Initially, ray tracing techniques were developed to describe mathematically radiant energy exchange between simply arranged surfaces. The techniques utilized the idea of following the history of radiation emitted from one surface and reflected or absorbed by others. They proved to be effective in finding closed form solutions to resulting infinite series (describing multiplicity of reflection and absorption) only in relatively simple surface configurations. However, when many surfaces were considered or an attenuation medium was present between the bounding surfaces, the original ray-tracing techniques failed.

In the 1930s the pioneering work by Hottel [3, 4] provided a methodology for calculating radiant exchanges in multi-surface configuration applications. The absence of high-speed computational capability caused Hottel to develop methods that were amenable to hand calculations or that made reference to auxiliary tools such as graphs. These graphs were constructed from data that were painstaking measured and extrapolated to useful parameter ranges.

The major breakthrough in radiative heat transfer calculations in the 1940s was Hottel's treatment of an absorbing isothermal gas in enclosures using the mean beam length approximation. Further extensions of this approach culminated in the zoning method, which allowed consideration of non-uniform gas temperatures.

During the 1950s, innovative applications of thermal radiation principles outside conventional power plant and furnace engineering put increasing demands on the understanding of the mechanism and the accuracy of radiation calculations. In response to this need an International Flame Radiation Research Committee, which in 1955 transformed into the present International Flame Research Foundation, was set up to sponsor research on flame radiation in furnaces [5]-[8].

The intensified interest and the need for greater sophistication in the treatment of engineering radiative transfer encouraged the publication of a number of textbooks devoted to the subject during the 1960s. Throughout the 1970s and up to now, the search has continued for accurate methods that could be broadly applied to multidimensional enclosures, with the additional capability of handling spectral characteristics of participating media.

1.3   Motivations, Objectives and Scope

Nowadays, with the dawning of the 21st century, the control of heat transfer mechanism inside enclosures, and particularly radiant energy transfer, cannot be based on a black-box principle fed with distributions of temperature and gas species concentrations. Solving the radiation problem requires, above all insight, an understanding of the realm of the process. Such understanding comes not only from experiments but also from the ability to see the problem in a wider theoretical context.

1.3.1   Motivations and Objectives

Numerically, an immense number of so-called *exact solutions* for radiation heat transfer have been accumulated in the literature over the past years only for very simple cases. All of these numerical solutions assume that the properties of the participating media are spatially uniform and the interior surfaces of the enclosure are radiatively black [9]-[11].

In engineering problems, however, radiative heat transfer calculations are very complex due to physical phenomena taking place within enclosures of complex shape. Furthermore, radiation problems are formulated mathematically as integral equations. Because of its importance, an exact mathematical solution to radiation exchange problems in enclosures is desirable. It is therefore of primary importance to develop an efficient and accurate method of solution which will ease the computational effort involved in radiation and in general heat transfer problems. The present work is designed to serve this purpose. As pointed out in a very significant paper by Howell [12]:

"... the technique must be applied in multi-dimensional problems, work well if the properties of the participating medium are nonuniform in space, lend itself to a match with the girding requirements of related conduction or convection equations, easily incorporate spectral properties, and treat anisotropic scattering. If not, the study of another one dimensional gray gas solution is pointless, even if the mathematical technique is elegant. No one will bother to use elegant solutions that cannot be applied to problems of technical interest."

Often for studies in radiation transfer, engineers went ahead and designed numerical methods for radiant heat exchange calculations. They did not bother whether their problems could be treated at best if they addressed the question of existence of a solution at the early stage. This rift has weakened fundamental developments in radiation transfer. It has also increased the uncertainty in predicting radiant heat exchanges in enclosures, as illuminated in an article by Tong and Skocypec [13]. Thus, the motivations for the present study are numerous.

From the mathematical view point, there is:

- A strong desire to define a mathematical framework which will provide the theory with the error analysis for heat transfer by radiation;

- A need to prove the solvability (existence and uniqueness of the solution) of the radiation heat transfer equation;

- A need to develop an exact mathematical algorithm for solution of the radiative transfer equation for any temperature distribution in an enclosure of arbitrary shape.

From the engineering view point, there is:

- A demand for a new, accurate and robust predictive procedure for computing radiant energy transfer in both flames and furnaces with a higher level of confidence [13];

- A demand for a fast method compatible with algorithms employed to solve the energy conservation equation.

With the foregoing introductory remarks in mind, we may rephrase the main objective of this book as to provide a unified and reliable treatment of the solution to radiation heat transfer problem in enclosures.

From the mathematical point of view, the stress is placed on:
- Existence and uniqueness of the theoretical solution;
- Existence and uniqueness of the associated numerical approximation;
- Accuracy to which numerical values are to be given.

From the radiative heat transfer point of view, the method is expected to:
- Be used in multi-dimensional (2D or 3D) enclosure geometries;
- Allow for non-uniformity of temperature distribution;
- Allow for special non-uniformity in temperature and gas properties;
- Allow for spectral (wavelength or frequency) calculations;
- Allow for anisotropic scattering;
- Allow for compatibility with algorithms that perform an energy balance.

Instead of starting with a simplified physical picture, we shall begin with a precise formulation of the most general problem. Then, subsequent simplified cases may be derived from the most general problem. To obtain a useful approximate solution we shall show that the answer to our problem can also be derived while considering the solution of a somewhat different problem obtained by using the variational principle.

1.3.2 Scope Covered

Many problems of mathematical physics are associated with the calculus of variations, one of the central fields of numerical analysis. The importance of a combined use of variational formulations and finite element approximation in many areas of applied science has already been demonstrated. An important advantage of this approach arises from the fact that the variational equations give stable numerical approximations. Here is meant that a variation of the given data in a sufficiently small range leads to an arbitrary small change in the numerical representation. We recall that in mathematical physics, a problem cannot be considered as realistically corresponding to physical phenomena unless this condition is established.

The characteristic difficulty of the variational approach, however, is that problems which can be meaningfully formulated may not have solutions.

If a variational principle for radiation heat transfer can be formulated on a well-specified set of admissible functions, then means are immediately established for obtaining numerical solutions in the standard integral form already available for the finite element analysis. Variational methods in radiation heat transfer analysis have been applied to the two parallel-plates system [14]-[16].

This treatise deals with the most general situation in which radiative properties vary with wavelength, temperature and medium composition. We restrict ourselves to three dimensions enclosure geometries without affecting the applicability of our results to two dimensions. Also we shall be concerned here with many topics of mathematical analysis related to boundary value problems. These include: the fundamentals of partial differential equations [17, 18, 20, 21]; the analysis of integral equations [17],[19]-[26]; the Lebesgue measure and integration theory [27]-[30]; the approximation analysis and the theory of ideal functions which, introduced by S L Sobolev [31] and latter expounded and called "distributions" by Laurent Schwartz [32], have become an indispensable tool for advanced calculus [19],[33]-[35]; and the fundamentals of variational calculus[36, 37]. Furthermore, results of functional analysis on Hilbert and Sobolev spaces [19]-[38] and spectral theory (eigenvalues and eigenfunctions) [18, 19, 39] are applied throughout the book.

The next section outlines the basis concepts of boundary value problems and the general line of approach adopted for their numerical solutions.

1.4   Basic Concepts in Boundary Value Problems

We limit ourselves in this section to a brief presentation of the main points concerning boundary value problems and the construction of the associated numerical solutions.

Let D be a specified region in the finite dimensional space $\mathbb{R}^n$ ($n = 2, 3$). Our discussion here merely concern problems of mathematical physics which are formulated in the form:

$$F(u(p), \partial^\alpha u(p)) = g(p), \qquad p \in D \qquad (1.1)$$

with the subsidiary boundary condition

$$B(u(r), \partial^\beta u(r)) = b(r), \qquad r \in \partial D \qquad (1.2)$$

Here, F is a function of the variables $u(p)$ and $\partial^\alpha u(p)$; that is, an operator which transform the functions $u(p)$ and $\partial^\alpha u(p)$ into another function $F(u(p), \partial^\alpha u(p))$. B is a function of the variables $u(r)$ and $\partial^\beta u(r)$, α and β are positive numbers. The functions $g(p)$ and $b(r)$ are prescribed physical data, and the function $u(p)$ of the independent variable p is sought such that (1.1) is identically satisfied in this variable if $u(p)$ and all derivatives $\partial^\alpha u(p)$ of relevant orders are substituted in F. We shall take the liberty of denoting by $S = \partial D$ the boundary of the fundamental domain D.

1.4.1 Typical Examples

A typical problem is to find a solution u to the equation

$$\Delta u(p) = g(p), \qquad p \in D$$

which is regular in the interior of the domain D, i.e. continuous there together with its first and second derivatives. The symbol Δ indicates the Laplace differential operator. Here, the function g is defined and possess continuous first derivatives for all $p \in D$ and it is bounded:

$$|g(p)| \leq N \ .$$

Another illustrative example, the generalized Stoke's problem, is to find a solution u to the equation

$$\alpha \cdot u(p) + \beta \cdot \Delta u(p) = g(p), \qquad p \in D$$

which is regular in the interior of the domain D, i.e. continuous there together with its first and second derivatives. The function g is defined and possess continuous first derivatives for all $p \in D$ and is bounded.

This problem is of significant importance in solving Navier–Stokes equations for flows with low Reynolds number.

Unless a set of subsidiary conditions are prescribed at the boundary ∂D of the domain, as shown in Fig. 1.2, the previous differential equations will have numerous solutions. The relations that described the boundary conditions may be linear or non linear. In practice, most of the questions dealt with are concerned with linear boundary conditions. These conditions are often classified into the following three categories:

Boundary Conditions of the First Kind

We speak of a boundary condition of the first kind when the sought for function u is prescribed at the boundary ∂D; that is,

$$u(r) = b(r), \qquad r \in \partial D$$

and we have the special case called *homogeneous boundary condition of the first kind* if the function u vanishes at the boundary ∂D.

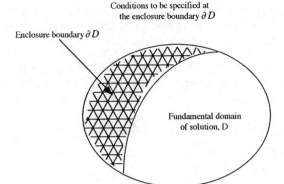

Figure 1.2. Schematic of the domain of interest and its boundary conditions

Boundary Conditions of the Second Kind

We speak of a boundary condition of the second kind when the normal derivative of the sought for function u is prescribed at the boundary; that is,

$$\frac{\partial u(r)}{\partial n(r)} = b(r), \qquad r \in \partial D$$

where $\partial/\partial n(r)$ denotes differentiation along the inner/outer normal $n(r)$ at the point r of the boundary. We have the special case called *homogeneous boundary condition of the second kind* if the normal derivative of the function u vanishes at the boundary ∂D.

Boundary Conditions of the Third Kind

We speak of a boundary condition of the third kind when a linear combination of the sought for function u and its normal derivative is prescribed at the boundary; that is,

$$\alpha \cdot u(r) + \beta \cdot \frac{\partial u(r)}{\partial n(r)} = b(r), \qquad r \in \partial D.$$

The boundary conditions of the first and second kind discussed already are obtained by choosing α and β equal to zero respectively.

For problems where the sought for function u represents a temperature distribution, the physical significance of the linear relation above is that heat is either removed or supplied by convection at the boundary ∂D.

The special case

$$\alpha \cdot u(r) + \beta \cdot \frac{\partial u(r)}{\partial n(r)} = 0, \qquad r \in \partial D$$

is called *homogeneous boundary condition of the third kind*.

These three types of linear boundary conditions cover most cases of practical interest. One, two or all of these conditions may be prescribed at the boundary of the domain under consideration.

There are also non linear boundary conditions such as the natural convection boundary condition and the thermal radiation boundary condition.

1.4.2 Integral-Transform Techniques

In solving the physical problem governed by (1.1), we seek to find an inverse operator F_α^{-1} such that

$$u(p) = F_\alpha^{-1}(g(p)), \qquad p \in D + S \qquad (1.3)$$

where g is arbitrarily given in a suitable restricted set of functions (e.g. g is continuous in $D + S$).

The operator F transforms a certain set of functions, in which it is applicable, into a wider function space. Conversely, the inverse transformation F_α^{-1} transforms the wider function space into a more restricted set of functions. In this sense it is often regarded as a smoothing transformation.

The smoothing transformation F_α^{-1} is in principle easy to handle and it is commonly represented by means of integral operators. Its determination is a matter of applying suitable analytic techniques which consist of rewriting (1.1). For a given problem, however, the choice of the technique depends on both the type of boundary conditions and the range of the space variables. These space variables can be finite, semi-finite, or infinitely extended.

Applying a specific analytic technique (e.g. Green's function, Green's integration theorem, Betti reciprocity theorem, etc.), the problem governed by (1.1) and (1.2) is reduced into an interrogation which is formulated at the boundary $S = \partial D$ and governed by an equation written in the form:

$$A(u(r)) = f(r) , \qquad r \in S. \qquad (1.4)$$

Here, A is either a linear integral operator, an integro-differential operator, or a pseudo-differential operator of a given order; and the function f provides a description of the physical situation over the entire domain of interest including its boundary.

Whilst finding a candidate u for which

$$
\begin{aligned}
F(u(p), \partial^\alpha u(p)) &= g(p), & p \in D \\
B(u(r), \partial^\beta u(r)) &= b(r), & r \in S
\end{aligned}
$$

aims at a solution in the full fundamental domain D, the alternative formulated merely at the boundary by (1.4) presents a much simpler problem.

Therefore, we shall avail ourselves with the question of finding a solution u such that the relation

$$A(u(r)) = f(r), \qquad r \in S$$

is valid under suitable subsidiary assumptions.

It is this boundary value problem combined with the finite element approximation technique which lead to the so called *Boundary Element Method - BEM -* in engineering analysis.

This line of approach proves to be very efficient since the original problem in the fundamental domain $D \subset \mathbb{R}^n$ is reduced to a problem in the $(n-1)$-dimensional compact manifold $S \subset \mathbb{R}^{n-1}$. From a numerical view point, it allows for the use of two different partitions: a partition (relatively fine) of the domain $D \subset \mathbb{R}^n$ for approximation of the physical quantities involved within the domain, and a partition (relatively coarse) of the boundary $S \subset \mathbb{R}^{n-1}$ for approximation of the solution and the physical quantities involved at the boundary. In both cases finite elements are used for numerical analysis. As such the numerical method remains compatible with algorithms employed to solved questions formulated within the fundamental domain D.

This method of solution, the BEM, does have a wide range of applications in engineering practice, and is of increasing importance. Problems governed by Laplace and Poisson's equations, elasticity problems, and problems in electrostatics and in the field of electromagnetic, which have been treated by means of boundary element method [40]-[44]. Here, the mathematical foundation for rigorous error and convergence analysis is provided by the variational or weak formulation of the boundary value problem.

The variational approach in engineering analysis is also called weighted residual technique. An adequate mathematical representation of the physical reality issued from the variational formulation often requires admission of ideal functions also called distributions. These ideal functions are basic in the study of variational formulations. They are introduced in order to widen the scope of elementary linear calculus by removing the straitjacket of differentiability and/or integrability conditions.

A coherent general theory of distributions can be found in the encyclopedia by Dautray and Lions [19]. We shall use them systematically in this book.

We may introduce distributions by strong definitions through completion of dense sets of smooth functions by convergence in a quadratic Hilbert norm. Such a procedure underlines the basic operations used in the variational theory.

We may also introduce distributions by means of weak definition; that is, instead of characterizing a continuous function f, defined on an $(n-1)$-manifold S, by the store of its values we characterize the function f weakly by the totality of inner products (f, φ) with all admissible functions φ of a class of functions whose support is in S. In other words, we consider distributions as elements in the dual space just as in projective geometry planes and points correspond dually to each other with respect to the inner product of their coordinates.

More explicitly, Let D be a specified region in the finite dimensional space $\mathbb{R}^n$ ($n = 2, 3$). If the boundary value problem (1.4) is to be solved on the boundary $S = \partial D$ under suitable subsidiary assumptions, then the corresponding weak formulation reads as to:

Find a candidate u such that for all admissible functions φ in an appropriated Hilbert space, there holds on S the relation

$$(A(u), \varphi) = (f, \varphi) .$$

The inner product (f, g) of two real-valued functions f and g is, of course, defined by the self-explanatory relation

$$(f, g) = \int_S f(r)g(r)dS(r) .$$

Thus, the variational problem is concerned with the question of determining a function $u \in U(S)$ such that for all admissible functions $\varphi \in H(S)$ the relation

$$\int_S A(u(r))\varphi(r)dS(r) = \int_S f(r)\varphi(r)dS(r) \tag{1.5}$$

is satisfied.

Whether or not one can retrieve the solution u, in the strong sense, of the boundary value problem (1.4) from the above variational equation is another story. Yet fortunately, difficulties arising from analysis of the variational formulation do not in general interfere with uniqueness and existence proofs of the solution of the boundary value problem. Knowing the properties of the linear operator A and having at our disposal the fundamentals of variational calculus we can identify a number of features of the solution u. Furthermore, a numerical algorithm whose skeleton is presented in the following paragraph can be constructed.

1.4.3 Numerical Approximations

The finite element approximation provides a standard methodology for the numerical treatment of the variational formulation governed by (1.5). To be specific, the numerical problem based on finite element approximation for the variational formulation consists of determining a solution u_h in a certain finite-dimensional subspace $U_h(S)$ of $U(S)$ such that the relation

$$\int_S A(u_h(r))\varphi_h(r)dS(r) = \int_S f(r)\varphi_h(r)dS(r)$$

remains valid for all admissible functions φ_h in an appropriated finite-dimensional subspace $H_h(S)$ of $H(S)$.

Here, we assign the subscript h to these subspaces to imply that their properties generally depend on a size parameter h (e.g. a mesh size or a maximum mesh diameter). As the size parameter h decreases, the dimensions $M_h = \dim(U_h(S))$ of $U_h(S)$ and $N_h = \dim(U_h(S))$ of $H_h(S)$ should increase and each subspace must tend to fill up $U(S)$ and $H(S)$ respectively.

The quality of the numerical approximation u_h of u, hence the very existence of u_h, depends on the properties of the subspaces $U_h(S)$ and $H_h(S)$. This is where the finite element approximation comes into view and the task is to precisely determine and build up these subspaces.

The construction of the subspaces $U_h(S)$ and $H_h(S)$ proceeds as follows. First, we imagine a partition S_h of finite surface elements covering the boundary S, and we approximate S by S_h using local geometrical interpolation upon fitting finite surface elements together to form a connected model. Next, we determine sets of global polynomial basis functions (shape functions)

$$\{\phi_i;\ 1 \le i \le \dim(U_h(S))\}\ of\ U_h(S)$$

and

$$\{\psi_i;\ 1 \le i \le \dim(H_h(S))\}\ of\ H_h(S),$$

and we expand the numerical solution u_h and the admissible functions φ_h as the linear combinations

$$u_h(r) = u_{h,1}\phi_1(r) + u_{h,2}\phi_2(r) + \cdots + u_{h,M_h}\phi_{M_h}(r)\ ,$$
$$\varphi_h(r) = \varphi_{h,1}\psi_1(r) + \varphi_{h,2}\psi_2(r) + \cdots + \varphi_{h,N_h}\psi_{N_h}(r)\ .$$

Then, the numerical problem becomes a question of finding the specific coefficients $u_{h,1}, u_{h,2}, \cdots, u_{h,M_h}$ which determine the solution of a variational system of equations.

The choice of both the interpolation functions and the basis functions ϕ_i and ψ_i, and their complexity, depend on the physical problem. Still, intuition is needed to make an appropriate choice.

The set $U(S)$ is often referred to as the space of trial functions, and $H(S)$ is referred to as the space of test functions.

When $U(S) = H(S)$, and we usually take $U_h(S) = H_h(S)$, the numerical method is known as the Rayleigh–Ritz or Galerkin technique and it yields a symmetric system of variational equations. These symmetric equations have emerged as a central and most fruitful topic in numerical analysis. In the context of mathematical physics nonsymmetric equations are of secondary importance.

When the space $H(S)$ contains all delta Dirac distributions, the equation governing the boundary value problem must be satisfied all over the domain

S. In this case, we take $U_h(S) \neq H_h(S)$ and let $H_h(S)$ be the space of N_h delta Dirac distributions

$$\psi_i(r) = \delta(r - r_i), \quad i = 1, \cdots, N_h .$$

Then we ask for solutions u_h in the subspace $U_h(S)$ such that

$$\int_S A\left(u_h\left(r\right)\right)\psi_i\left(r\right)dS\left(r\right) = \int_S f\left(r\right)\psi_i\left(r\right)dS\left(r\right) .$$

Using the definition of delta Dirac distributions, we immediately obtain the relations

$$A\left(u_h\left(r_i\right)\right) = f\left(r_i\right), \qquad r_i \in S = \partial D, \quad i = 1, \cdots, N_h .$$

These relations traduce the fact that the numerical solution u_h must satisfy the governing equation of the boundary value problem at the N_h chosen collocation points. As a consequence, this numerical method is often referred to as the points collocation technique or the Nyström technique. It has recently been used in the numerical solution of complex radiation problems [45].

2
Physical Model

In this chapter, we shall examine the physical model of radiative heat transfer and its basic laws. The chapter begins with a description and definition of physical quantities arising in radiation and is concerned with the mathematical relations describing radiative heat transfer.

2.1 Emitted Radiation

All substances continuously emit radiation by virtue of the molecular and atomic agitation associated with the internal energy of the material. Quantum theory views radiation as the propagation of photons. Alternatively, radiation may be viewed as the propagation of electromagnetic waves according to the description of electromagnetic theory.

Physically, if we conceive radiation as the propagation of photons then, in media with constant refractive index, this transport takes place along straight lines referred to as lines of sight. Since photons move with a velocity which is of the same order as the speed of light in vacuum, steady state is reached almost instantaneously so that transient states of radiation may nearly always be neglected in engineering applications.

The fundamental physical quantity that characterizes radiation is the photon density. We shall define the photon density as the number of photons per unit volume, moving forward within a given solid angle, and in a given wavelength interval. The spectral intensity, which is the amount of energy transported, is often used as fundamental quantity. To introduce this notion we consider a pencil of ray originating from a point having a vector coordinate r to a point p.

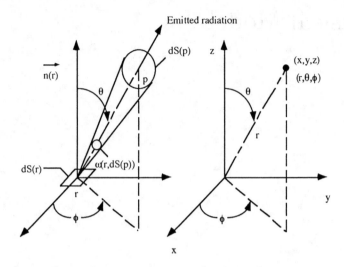

Figure 2.1. Directional nature and notation used in the definition of intensity of emitted radiation

As shown in Fig. 2.1, the direction of travel $\overrightarrow{rp}$ may be specified in terms of zenith (or polar) and azimuth angles, $\theta(p, r)$ and $\phi(p, r)$ respectively, of a spherical coordinate system. We consider a differential small surface $dS(p)$ at the point p and normal to the direction $\overrightarrow{rp}$. The surface $dS(p)$ subtends a solid angle $\omega(r, dS(p))$ at the point r.

2.1.1 Intensity of Emitted Radiation

The spectral intensity $I^e(r, p, \lambda)$ of emitted radiation from the point r into the direction $\overrightarrow{rp}$ is formally defined as the rate at which radiant energy is emitted from r at the wavelength λ in the direction $\overrightarrow{rp}$, per unit area normal to this direction, per unit solid angle about this direction, and per unit wavelength interval about λ. Accordingly, the following applies:

$$I^e(r, p, \lambda) = \frac{\partial^3 E(r, p, \lambda)}{\partial(S(p)) \partial \omega(r, dS(p)) \partial \lambda} ,$$

with $E(r, p, \lambda)$ being the energy of emitted radiation per unit of time.

Mathematically we have expressed the spectral intensity $I^e(r, p, \lambda)$ in terms of spatial positions r and p, and wavelength λ. The objective in radiative transfer theory is to determine $I^e(r, p, \lambda)$ as a function of these independent variables.

The spectral nature of intensity is one of the two features that complicates the radiative heat transfer analysis. The second feature relates to its directional dependence; the direction being specified here by the points r and p. An infinite number of independent values of intensity exist at every point of the medium traversed by radiation. To properly quantify radiant energy exchange, we must be able to treat both the spectral and directional effects.

The total intensity of emitted radiation at any point in the enclosure is obtained by integrating the spectral intensity $I^e(r, p, \lambda)$; that is,

$$I^e(r, p) = \int_0^{+\infty} I^e(r, p, \lambda) d\lambda .$$

For gas-filled enclosures, which is usually the case in industrial furnaces or combustion chambers, the medium and the enclosing surface properties vary so irregularly with wavelength that detailed integration over wavelengths is a practical impossibility. However, combustion chambers in industry are often filled in with heteropolar gases and gases with symmetric diatomic molecules.

Heteropolar gases such as H_2O, CO_2, CO, NO and various hydrocarbons radiate to an appreciable extent at given wavelengths. These gases are asymmetric in one or more of their modes of vibrations. During molecular collisions, rotation and vibrations, individual atoms in a molecule can be excited so that atoms that possess free electrical charges can emit electromagnetic waves. When radiation of the appropriate wavelength impinges on such gases, it is enhanced in the process.

Gases with symmetric diatomic molecules such as N_2, O_2, H_2, dry air and others nonpolar symmetrical molecular structure neither emit nor absorb radiation, unless heated to such extremely high temperatures that they become ionized plasmas and electronic energy transformations occur.

These observations lead to the consideration of finite wavelength bands in gas-filled furnaces. Wavelength bands are hypothetical models of simplified mathematical structure which are introduced to provide fair representations of the properties of real gas spectra at a reasonable computing cost.

In wavelength band considerations, the whole spectrum is divided into absorbing and non-absorbing bands as illustrated in Fig. 2.2. The non-absorbing bands are also called windows of the spectrum. The spectral intensity of emitted radiation is determined for each band and the total intensity $I^e(r, p)$ is obtained by summing up over the entire spectrum the band contributions $I^e(r, p, \Delta\lambda)$ defined as:

$$I^e(r, p, \Delta\lambda) = \int_{\Delta\lambda} I^e(r, p, \lambda) d\lambda .$$

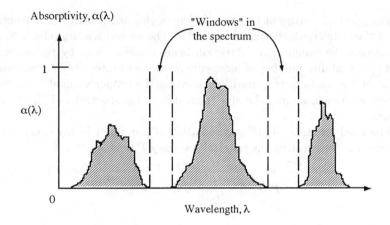

Figure 2.2. Variation of gas absorptivity with wavelength

Now, we shall explicit the laws relating radiative heat transfer to the temperature of a matter.

2.1.2 Radiation from a Blackbody

When radiant energy is emitted by the ideal material, termed blackbody, the emitted intensity is independent of direction. It will be given by the distribution $I_b(r, \lambda)$. This concept of blackbody is fundamental to the study of radiative heat transfer.

We merely recall that a blackbody is a perfect absorber and emitter of energy. It provides a suitable standard with which real absorber, hence real emitter of energy can be compared. Only a few materials such as carbon black, carborundum, platinum black, and gold black approach the blackbody in their ability to absorb radiant energy.

For a blackbody, the intensity of emitted radiation from point r at temperature $T(r)$ in a medium with constant refractive index n is defined to be:

$$I_b(r, \lambda) = \frac{C_1/\pi}{n^2 \lambda^5 \left(\exp\left(\frac{C_2}{n\lambda T(r)}\right) - 1\right)}.$$

This relation remains valid for emission into a medium whose index of absorption is much less than the index of refraction. Henceforward, we can systematically account for the effect of index of refraction n on the intensity by adopting the function $n^{-2} \cdot I^e(r, p, \lambda)$ rather than $I^e(r, p, \lambda)$. Any changes subsequently observed in $n^{-2} \cdot I^e(r, p, \lambda)$ will then be caused solely by the process of absorption, emission and scattering. For most engineering applications, emission takes place into air or other gases with index of refraction n close to unity.

Table 2.1. Adjusted values of radiation constants [47]

Constant and symbol	Value	Unit (SI)
Speed of light in vacuum, c_o	2.997925E+08	m/s
First radiation constant, C_1	3.741844E-16	Wm^2
Second radiation constant, C_2	1.438833E-02	mK
Radiation constant, C_3	2.897000E-03	mK
Planck's constant, h	6.626196E-34	Js
Boltzmann constant, k	1.380622E-23	J/K
Stefan-Boltzmann constant, σ	5.669610E-08	$W/(mK)^2$

The following function

$$e_b(r, \lambda) = \pi \cdot I_b(r, \lambda)$$

is known as Planck's spectral distribution or emissive power. The terms $C_1 = 2\pi hc^2$ and $C_2 = hc/k$ are the first and second radiation constants respectively and their values are listed on Table 2.1, h is Planck's constant, k is the Boltzmann constant, and $c \equiv c_o$ is the speed of light in vacuum.

We can determine the wavelength, $\lambda_{\max}$, for which Planck's spectral distribution attains its maximum by differentiating the relation above and setting the result to zero. This lead to (Wien's displacement law):

$$\lambda_{\max}(T(r)) = \frac{C_3}{nT(r)} .$$

For a blackbody, the Stefan–Boltzmann law shows that in a medium with constant refractive index n the total emissive power within the entire spectrum is given by the relation:

$$e_b(r) = n^2\sigma \cdot (T(r))^4 ,$$

where $\sigma = (C_1/15) \cdot (\pi/C_2)^4$ is the Stefan–Boltzmann constant.

This is a very important result since it illustrates the fourth power temperature dependence for blackbody radiation, and establishes the fact that radiant energy transfer becomes increasingly important as the temperature increases.

In radiative heat exchange calculations it is often desirable to determine the power that is emitted in a given wavelength band as illustrated in Fig. 2.3. The emissive power $e_b(r, \Delta\lambda)$, at the point r of temperature $T(r)$, contained within the finite wavelength interval $\Delta\lambda = [\lambda_1, \lambda_2]$ is given by the following relation:

$$e_b(r, \Delta\lambda) = \int_{\Delta\lambda} e_b(r, \lambda)d\lambda ,$$

or equivalently,

$$e_b(r, \Delta\lambda) = n^2\sigma \cdot (T(r))^4 \left(f(n\lambda_1 T(r)) - f(n\lambda_2 T(r)) \right) ,$$

$$f(n\lambda T) = \frac{15}{\pi^4} \int\limits_{C_2/(n\lambda T)}^{+\infty} \frac{x^3}{e^x - 1} dx .$$

Approximation expressions for the function $f(n\lambda T)$ are [2]:

- for $x = \frac{C_2}{n\lambda T} \geq 2$

$$f(n\lambda T) = \frac{15}{\pi^4} \sum_{m=1}^{+\infty} \frac{1}{m^4 e^{mx}} \left\{ [(mx + 3)mx + 6] \, mx + 6 \right\},$$

- for $x = \frac{C_2}{n\lambda T} < 2$

$$f(n\lambda T) = 1 - \frac{15}{\pi^4} x^3 \left(\frac{1}{3} - \frac{x}{8} + \frac{x^2}{60} - \frac{x^4}{5040} + \frac{x^6}{272160} - \frac{x^8}{13305600} \right).$$

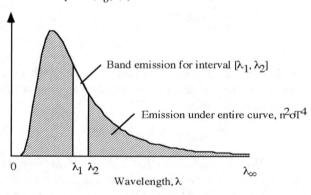

Spectral emissive power, $e_b(T,\lambda)$

Band emission for interval $[\lambda_1, \lambda_2]$

Emission under entire curve, $n^2\sigma T^4$

Wavelength, λ

Figure 2.3. Spectral blackbody emissive power for a given temperature

Once the total intensity of emitted radiation is known, the flux of emitted radiation is determined by integrating the intensity field over all solid angles $\omega(r, dS(p))$.

Merely for the sake of compactness we shall omit the explicit use of the wavelength variable λ. It will be used explicitly only when integration over the entire spectrum is required. We shall also restrict ourselves to media with index of refraction close to unity.

2.2   Incident, Absorbed and Scattered Radiation

The foregoing concepts of radiation emitted by a matter can be extended to incident radiation. As shown in Fig. 2.4, such radiation originates from emissions and reflections occurring at the enclosure boundary. It will have directional distribution determined by the intensity $I^i(r,p)$. The intensity $I^i(r,p)$ is defined as the rate at which radiant energy is incident at point p from direction $\overrightarrow{rp}$, per unit area of intercepting surface normal to this direction, and per unit solid angle $\omega(p, dS(r))$.

We begin by considering a pencil of incident radiation of intensity $I^i(r,u)$ which propagates within a region of absorbing, emitting and scattering medium as shown in Fig. 2.5. The change in intensity along a sufficiently small path $dL(u)$ is due to the processes of absorption, emission and scattering.

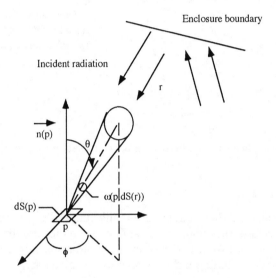

Figure 2.4. Directional nature of incident radiation

2.2.1   Absorption

We introduce a coefficient of proportionality $a(u)$ which depends only on the local properties of the medium. Then, the decrease of intensity due to absorption (including the contribution by induced emission) is given by:

$$\frac{dI^i(r,u)}{dL(u)}\bigg|_{\text{absorption}} = -a(u)I^i(r,u).$$

The coefficient $a(u)$ describes the amount of incident radiant energy absorbed per unit path length within the medium. It is a function of the state variables only; that is, the density and temperature, as well as the composition of the medium. The absorption coefficient $a(u)$ depends on the wavelength of radiation, but it does not depend on the direction (or path length) of radiation.

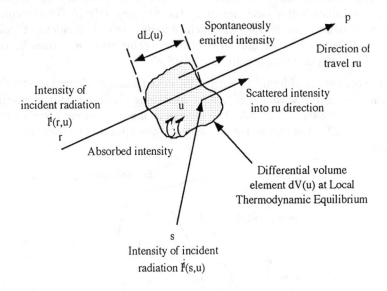

Figure 2.5. Geometry for derivation of the radiative transfer equation

2.2.2 Emission

We assume that local thermodynamic equilibrium (LTE) prevails inside the medium. Therefore, the spontaneous emission contribution by the medium along the path $dL(u)$ to intensity of incident radiation at the point u from direction $\overrightarrow{ru}$ (not including induced emission) is given as:

$$\left. \frac{dI^i(r,u)}{dL(u)} \right|_{\text{emission}} = a(u)I_b(u) \, .$$

The local thermodynamic equilibrium hypothesis implies the following three conditions:

- The system under study can be divided into elements small enough so that macroscopic thermodynamic properties can be assumed constant within each element;

- The fluctuation in the thermodynamic variables must remain small in each element if compared to their average value, furthermore, external field gradients must be such that induced relative changes in

thermodynamic variables over an element remain either less than or
of the order of the equilibrium fluctuation in that element;

- The time scale fluctuations of the thermodynamic properties control-
ling the process under consideration must be infinitely fast if compared to the rate of change of the external field properties.

2.2.3 Scattering

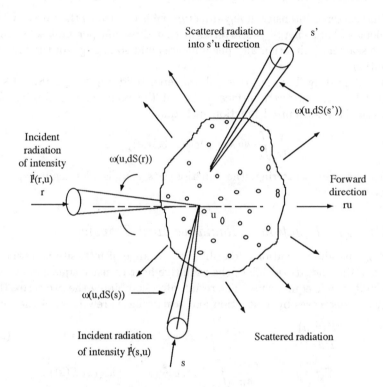

Figure 2.6. Scattering in a medium containing particulate matter

For, the scattering particles inside the medium randomly oriented so that
their scattering cross section is independent of the incidence direction, we
may write the sum of the attenuation by scattering and the gain by incoming scattering in the direction $\overrightarrow{ru}$ as:

$$\left.\frac{dI^i(r,u)}{dL(u)}\right|_{\text{scattering}} = -\sigma(u)I^i(r,u) + \Sigma(r,u) ,$$

with

$$\Sigma(r,u) = \frac{\sigma(u)}{4\pi} \int_{4\pi} I^i(r,u)\Phi(s,r,u)d\omega(u,dS(s)) .$$

Here, the coefficient $\sigma(u)$ describes the amount of incident radiant energy per unit path length for propagation in the medium owing to scattering phenomenon. This scattering coefficient is a function of the particle number density and the particle scattering cross section. The quantity $\Phi(s,r,u)$ is the dimensionless phase function for scattering. As illustrated in Fig. 2.6, the phase function $\Phi(s,r,u)$ has the physical interpretation of being the scattered intensity at the point u, of incident intensity from direction $\overrightarrow{su}$ into the direction $\overrightarrow{ru}$, divided by the intensity that would be scattered in that direction if the scattering was isotropic.

The scattered intensity in any direction $\overrightarrow{ru}$ is defined as the rate at which incident radiant energy is scattered in that direction per unit solid angle of the scattered direction and per unit area and solid angle of the incident radiation.

By integrating $\Phi(s,r,u)$ over all scattered solid angle $\omega(u, dS(s\prime))$ subtended by the scattered surface $dS(s\prime)$ at the point u, one obtains that $\Phi(s,r,u)$ is a normalized function such that:

$$\frac{1}{4\pi} \int_{4\pi} \Phi(s,r,u) d\omega(u, dS(s\prime)) = 1 \ .$$

For isotropic scattering, the function $\Phi(s,r,u)$ is identically equal to unity.

2.2.4   The Equation of Radiative Heat Transfer

Adding the above relations describing the change of intensity as radiation passes a distance $dL(u)$, gives the radiative heat transfer equation in local thermodynamic equilibrium. The result, after combining the two terms that represent decreases by absorption and scattering, is rewritten in the form:

$$\frac{dI^i(r,u)}{dL(u)} = -\beta(u)I^i(r,u) + a(u)I_b(u) + \Sigma(r,u) \ , \qquad (2.1)$$

$$\Sigma(r,u) = \frac{\sigma(u)}{4\pi} \int_{4\pi} I^i(r,u)\Phi(s,r,u) d\omega(u, dS(s)) \ ,$$

with $\beta(u) = a(u) + \sigma(u)$ being the medium extinction coefficient.

The above integro-differential relation is the radiative heat transfer equation. It describes the net rate of gain of intensity of an incident beam of radiation at each local point u as the beam travels along a path (r,p) throughout an absorbing-emitting and scattering medium.

We can observe that this equation devolves on the consideration of the appropriate description of the following two factors: the path (r,p) of an incident beam of radiation in an absorbing-emitting and scattering medium; the mathematical accounting for the net rate of change of intensity of incident radiation $I^i(r,u)$ describing the processes of absorption, emission and scattering.

2.2.5 Integral Form of Radiative Heat Transfer Equation

Formal integration of (2.1) along the path $(r, p) \ni u$ leads to the integrated directional equation, or integral form, of radiative heat transfer equation. The new equation reads:

$$I^i(r, p) = I^o(r, p)\tau(r, p) + \int_{(r,p)} \beta(u)G(r, u)\tau(u, p)dL(u) , \qquad (2.2)$$

where $I^o(r, p)$ is the intensity of emitted and reflected radiation at the origin point r into the direction of travel $\overrightarrow{rp}$, $\tau(r, p)$ is the transmissivity between the point r and p, and $G(r, u)$ is the source intensity along the path (r, u) from both emission and incoming scattering.

The source intensity $G(r, u)$ is given by the relation:

$$G(r, u) = \left(1 - \frac{\sigma(u)}{\beta(u)}\right) I_b(u) + \frac{\Sigma(r, u)}{\beta(u)} .$$

It defines the intensity in the direction of travel $\overrightarrow{ru}$ at the local position u arising from emission of radiation by the medium, plus the intensity scattered into the direction $\overrightarrow{ru}$ at the local position u due to incoming radiation from all 4π $(sr.)$ directions.

The transmissivity

$$\tau(r, p) = \exp\left(-\int_{(r,p)} \beta(u)dL(u)\right)$$

defines the fraction of incident radiation from the point r that penetrates through the path (r, p). If the extinction coefficient is constant, the transmissivity becomes

$$\tau(r, p) = \exp(-\beta \|r - p\|_{R^3}) .$$

2.3   Radiation from Particulate Matter

The intensity of radiation can be attenuated or enhanced by the process of scattering. Scattering is the redirection of radiation by interaction with molecules, particles or objects of any size and shape within the enclosure volume. The interaction can result from a combination of reflection, refraction, and diffraction as shown in Fig. 2.7. These complex interactions have been mathematically described by Gustav Mie who suggested a general treatment of light propagation problem in a turbid medium.

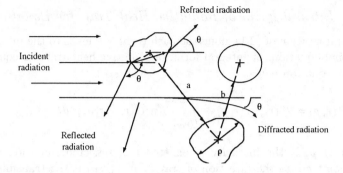

Figure 2.7. Scattering of radiation in a medium containing particulate matter

Here, we consider sperical particles and we view radiation as energy in transport in electromagnetic wave form with electric and magnetic vectors vibrating perpendicular to the direction of propagation. The treatment is applicable even for non-spherical particles that are randomly oriented. Indeed, since the particles are in a random orientation, an equal portion of surface element will face each angular direction. This portion receives the same angular distribution as for a spherical particle. Thus, the angular distribution of scattered radiation viewed at a distance from the actual particle will be the same as that scattered from spherical particles.

The Mie scattering theory is based on the solution of Maxwell equations for an incident beam of unpolarized radiation, in terms of scatter characteristics of a single particle [46]. Among these are the efficiency factors for extinction $E(m(u), x(u))$ and scattering $S(m(u), x(u))$. They are function of the complex index of refraction $m(u)$ and the size parameter $x(u)$ of a particle at the point u. Namely,

$$E(m, x) = \frac{2}{x^2} \sum_{n=1}^{+\infty} (2n + 1) \operatorname{Re}(a_n(m, x) + b_n(m, x)) ,$$

$$S(m, x) = \frac{2}{x^2} \sum_{n=1}^{+\infty} (2n + 1) \left(|a_n(m, x)|^2 + |b_n(m, x)|^2 \right) .$$

The efficiency factor for absorption $A(m, x)$ is given by the difference

$$A(m, x) = E(m, x) - S(m, x).$$

The size parameter $x \equiv x(u)$ of a particle of diameter $\rho(u)$ at the point u, interacting with a train of photons or electromagnetic waves at the wavelength λ, is defined to be:

$$x(u) = \frac{2\pi\rho(u)}{\lambda} .$$

The Mie scattering coefficients $a_n(m, x)$ and $b_n(m, x)$ are complex-valued functions defined by the relations:

$$a_n(m, x) = \frac{\psi_n'(mx)\psi_n(x) - m\psi_n(mx)\psi_n'(x)}{\psi_n'(mx)\zeta_n(x) - m\psi_n(mx)\zeta_n'(x)},$$

$$b_n(m, x) = \frac{m\psi_n'(mx)\psi_n(x) - \psi_n(mx)\psi_n'(x)}{m\psi_n'(mx)\zeta_n(x) - \psi_n(mx)\zeta_n'(x)}.$$

In these relations, the functions $\psi_n(z)$ and $\zeta_n(z)$ are the Riccati–Bessel functions and their derivatives with respect to the variable z are denoted by $\psi_n'(z)$ and $\zeta_n'(z)$. They can be calculated from the following recurrence relations [47]:

$$\psi_{n+1}(z) = \frac{2n+1}{z}\psi_n(z) - \psi_{n\ 1}(z),$$

$$\xi_{n+1}(z) = \frac{2n+1}{z}\xi_n(z) - \xi_{n\ 1}(z),$$

with

$$\zeta_n(z) = \psi_n(z) - i\xi_n(z),$$
$$\psi_0(z) = \sin(z),\quad \xi_0(z) = \cos(z);,$$
$$\psi_1(z) = \frac{1}{z}\sin(z) - \cos(z);,$$
$$\xi_1(z) = \sin(z) + \frac{1}{z}\cos(z).$$

If we assume an absence of interference between the different particles within the medium, we may determine the scattering coefficient $\sigma(u)$, of a particle at the point u in a particle cloud, by integrating the scattering cross section

$$\pi \cdot (\rho(u))^2 \cdot S(m(u), x(u))$$

over all particle diameters. Accordingly,

$$\sigma(u) = \pi \int_0^{+\infty} (\rho(u))^2 S(m(u), x(u)) f(\rho(u)) d\rho(u),$$

with $f(\rho)$ being the particles size distribution function.

In an analogous manner we can also define the particle extinction, hence absorption coefficient.

Now, let $\theta(s, r, u)$ designate the angle by which incident radiation from direction $\vec{su}$ is scattered into the direction $\vec{ru}$. Then, the phase function for scattering $\Phi(s, r, u)$ is defined to be:

$$\Phi(s, r, u) = \frac{2\left(|I_\perp(m, x, \theta(s, r, u))|^2 + |I_\parallel(m, x, \theta(s, r, u))|^2\right)}{x^2 S(m, x)}.$$

Here, $m \equiv m(u)$ and $x \equiv x(u)$. The complex-valued functions $I_\perp(m, x, \theta)$ and $I_{||}(m, x, \theta)$ appearing in the relation above are known as the amplitude functions. They are given by the relations:

$$I_\perp(m, x, \theta) = \sum_{n=1}^{+\infty} \frac{2n+1}{n(n+1)} \left(a_n(m, x) \frac{P_{n,1}(z)}{\sin(\theta)} + b_n(m, x) \frac{dP_{n,1}(z)}{d\theta} \right),$$

$$I_{||}(m, x, \theta) = \sum_{n=1}^{+\infty} \frac{2n+1}{n(n+1)} \left(a_n(m, x) \frac{dP_{n,1}(z)}{d\theta} + b_n(m, x) \frac{P_{n,1}(z)}{\sin(\theta)} \right),$$

with $P_{n,1}(z)$ being the associated Legendre functions of the first kind. The functions $P_{n,1}(z)$ are finite series defined for $z = \cos(\theta) \in [-1, 1]$ by the relation:

$$P_{n,1}(z) = \sqrt{1-z^2} \frac{d}{dz} \left(\frac{1}{2^n n!} \frac{d^n(z^2-1)^n}{dz^n} \right).$$

The foregoing relations are as yet general and of a purely formal character. We refer the reader to the special textbook by Van de Hulst [46] for further details on the subject of particle scattering. From the developments given above, it seems reasonable to assume that knowledge of particle scattering properties and their concentration would place us well on our way toward having a complete solution to the radiative heat transfer problem in enclosures. Unfortunately, these properties are related to the fuel type and mixing histories of the combustion process and they are seldom known. As a consequence, this insufficiency of physical information relegates the prediction of the radiative heat exchanges to non-scattering systems.

In most engineering applications, the influence of scattering mechanism onto the overall radiative transfer is negligible. Furnaces fired with gaseous and liquid fuels allow for such simplifications even in the presence of soot. For a beam of radiation passing through a carrier gas containing both suspended soot and particle matter, calculations have shown that scattering for soot is negligible and scattering for large particles is forward directed [48]. In such systems, scattering can be ignored to the first approximation. Since the attenuation of radiation intensity in this instance obeys Bourguer's law with a particle mean specific absorption coefficient, the contribution of scattering to radiation can be accounted for by increasing only the absorption coefficient.

Accordingly, from now onwards attention will be placed on emitting-absorbing media. Scattering will be discussed again in Chap. 9. In a system where the effect of scattering on the intensity of radiation can be ignored, the integrated directional equation of radiative heat transfer is reduced to:

$$I^i(r, p) = I^o(r, p)\tau(r, p) + \int_{(r,p)} a(u)I_b(u)\tau(u, p)dL(u), \qquad (2.3)$$

with the transmissivity $\tau(r,p)$ defined as:

$$\tau(r,p) = \exp\left(-\int_{(r,p)} a(u)dL(u)\right) .$$

2.4 Governing Equations with Shadow Zones

A good model of radiative heat transfer in enclosures should allow for predicting the radiative heat exchanges in participating media with the presence of shadow zones. The shadow zones are due to the enclosure geometry (non-convex[1] enclosure geometry) or to the presence and orientation of heat sink. Mathematically, these zones can be detected using a characteristic function $\chi(r,p)$.

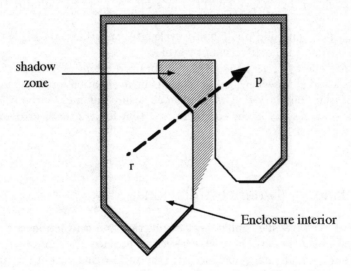

shadow zone

p

r

Enclosure interior

Figure 2.8. Enclosure with shadow zones

As can be seen in Fig. 2.8, the incoming intensity at any point p on a line of sight is equal to zero for a shadow zone lying between the origin r of that line and the observation point p, since this origin cannot be seen when looking from point p. The shadow zones can be detected directly by using an indicative function or, as will be shown latter in Sect. 2.6, they can be detected indirectly by including other properties in the formulation of the problem. A function identifying presence of shadow zones, termed the shadow zone function, is then defined as:

---

[1] We call an enclosure convex (curved outward) if it contains the straight segment joining any two points belonging to it.

$$\chi(r,p) = 1_{\Omega(p)}(r)\,,$$

with $\Omega(p)$ being the part of the enclosure interior surface which sees the point p; that is,

$$\Omega(p) = \{r \in S; \quad r \text{ is seen when looking from } p\}\,.$$

Therefore, the integrated directional equation of radiative heat transfer that takes into account the presence of shadow zones has the form:

$$I^i(r,p) = \left(I^o(r,p)\tau(r,p) + \int_{(r,p)} a(u)I_b(u)\tau(u,p)dL(u) \right) \chi(r,p).$$

The equation hereabove gives the intensity of incident radiation at the point p, originating from the location r, that is locally traveling in a single direction per unit solid angle and wavelength, and that is crossing a unit surface normal to the direction of travel.

The net energy crossing a surface is obtained upon integration over all directions and all wavelengths. This results in an equation for local radiative fluxes at the boundary and within the enclosure volume. Without loss of generality we assume in the considerations that follows that our domain is convex $\chi(r,p) \equiv 1$.

2.5   Energy Balance Relations

Radiative heat transfer, unlike heat conduction and convection, is an outstretched phenomenon. For domains such as furnaces, having opaque[2] enclosing surfaces, an energy balance over the surface and within the furnace interior must be performed. This section outlines the energy balance equations of radiative transfer that will be used to determine the intensity of radiation and the radiative fluxes in an absorbing-emitting medium.

As in other branches of engineering, the solution to the radiative heat transfer problem requires assumptions, idealizations and approximations. We shall first lay down engineering hypotheses and then proceed with some degree of confidence to the development of the solution to the radiative heat transfer in enclosures.

---

[2]participating surface on which radiation is absorbed in a very thin layer ($1\mu m$ to $1mm$) or in its vicinity; the participating surface is also thick enough that no electromagnetic waves can penetrate through it.

2.5.1 General Considerations and Boundary Conditions

The essential feature which differentiates engineering radiative heat transfer from astrophysics radiation is the inter reflections occuring at the enclosure boundary. In determining the radiative heat exchanges from a surface, it is necessary to possess information about all directions in space into which the surface may radiate or from which radiation may arrive.

Enclosure Considerations

To make certain that all radiation whether emitted or incident at a given location p is accounted for, first we figuratively construct an enclosure which encompasses all possible directions visible from its interior surface. The enclosure boundary need not necessarily consist of real surfaces (walls) only but it may include imaginary spaces as for example, a furnace exit. The exit will be treated as a surface of the enclosure. It will be a special surface which does not reflect any of the energy incident upon it. Additionally, it will be characterized by an effective radiant emissive power equal to the energy which passes into a blackbody cavity through the open end. Mathematically, the enclosure is a closed, connected, and bounded region D of the three dimensional world $\mathbb{R}^3$. We denote by $S = \partial D$ the boundary (i.e., the interior surface) of our enclosure. The enclosure interior $D - \partial D$ is the proper difference between the domain D and its bounding surface $S = \partial D$.

Lambert Boundary Conditions

Next, we shall assume that the interior surface of the enclosure is opaque and its directional property effects are independent of the intensity of incident and emitted radiation. We also assume that it can be treated as a Lambert surface; that is, a diffuse emitter, absorber and reflector of radiant energy. This latter assumption means that regardless of where the incident beam of radiation which impinges an arbitrary location p on the surface originates, the angular distribution of the reflected energy at the point p is uniform for all directions.

Thus, the previous history of the beam of incident radiation is completely obliterated when it impinges and when it is reflected at the surface. Since electromagnetic wave theory predicts a zero emissivity at a polar angle of 90° for all materials, no real surface can be a perfect diffuse emitter. However, experimental evidences, as shown in Fig. 2.9, demonstrate that most surfaces emit (and, therefore, absorb) diffusely except for polar angle θ greater than 60°. Furthermore, little energy is emitted by the surface into these directions. Thus, we can consider the assumption of diffuse emission as a good one.

Indeed, if enclosure boundary is made of electrical nonconductors or refractories we can assume that the directional nature of the properties at the boundary are independent of the direction for polar angles between $0°$ and $60°$ as shown in Fig. 2.9.

Now, let consider the radiant energy emitted by a blackbody within the cone bounded by the region $0 \leq \theta \leq \eta$ is given by $e_b(r)(1 - \cos^2(\eta))$. We find that 75% of the energy is emitted within the cone $0 \leq \theta \leq 60°$ and 97% within the cone $0 \leq \theta \leq 80°$.

Thus, it is acceptable for engineering purposes to assume that the enclosure boundary behaves like a Lambert surface. Besides, the error involved in measuring the material properties and the uncertainty involved in characterizing the state of the boundary are ambiguities larger than those introduced by the assumption of Lambert surfaces.

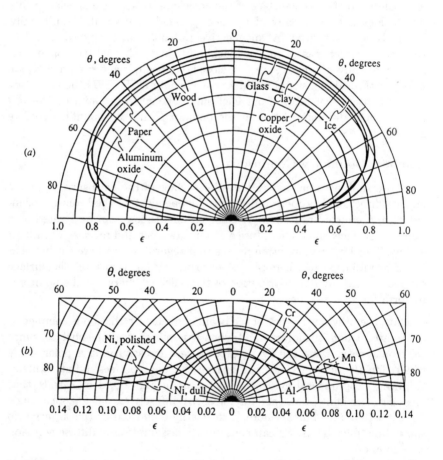

Figure 2.9. Total directional emissivities: a - for several nonconductors, b - for several metals [49]

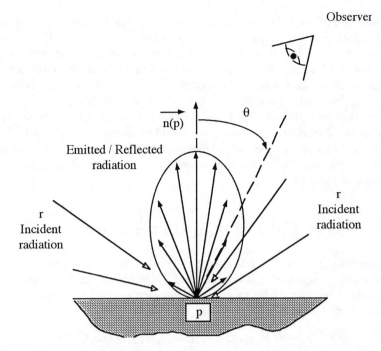

Figure 2.10. Directional distribution at the enclosure boundary

The utility of the Lambert assumption is that both emitted and reflected radiation at the interior surface have the same directional distribution. As shown in Fig. 2.10, this means that if an envelope is stretched over the surface S and an observer is stationed on the envelope watching energy from the surface pass by, he is unable to tell the difference between the emitted and reflected energy. Since this distinction cannot be made, it is convenient to express the intensity $I^o(r, p)$ of radiation leaving the location r as function of the variable r only, $I^o(r)$.

Extension of Prevost's Principle

Finally, we shall assume throughout the remaining of this book that the nature of radiation emitted at any point of the enclosure boundary depends only on the wavelength and on the thermodynamic and mechanical state of the boundary surface. For most cases this means dependence of the emitted radiation on temperature, surface roughness, and chemical composition. This is essentially an extension of Prevost's Principle which states that a body must continuously emit radiation even when it is in thermal equilibrium with its surroundings. We exclude from our considerations such phenomena as fluorescence, for there the emitted radiation from a surface depends on the incident radiation.

Thermal radiation is the propagation of photons. Photons essentially never undergo photon-photon collisions and the photon fluxes encounter in practice are so low that the probability of two photons arriving at the same absorption point is sufficiently small. Under these circumstances a photon striking the enclosure boundary surface will not be influenced by other photons striking the surface, nor will it be influenced by other photons emitted from the surface. Thus, the radiative properties of the material covering the enclosure boundary (i.e.; boundary surface emissivity and absorptivity[3]) can be regarded as independent of the intensity of incident and emitted radiation. Therefore, they depend only on the wavelength and on the state of the surface.

2.6   Energy Balance on a Unit Surface

The intensity of incident radiation is defined as the radiative energy flow per unit solid angle and unit area normal to the line of sight. It imparts an infinitesimal heat flow (energy absorbed) at any location p of the boundary surface in the amount of:

$$dQ^a(p) = \alpha(p)I^i(r,p)d(\text{proj}_{\perp(r,p)}(S(p)))d\omega(p,dS(r)) ,$$

or equivalently,

$$dQ^a(p) = \alpha(p)I^i(r,p)\cos(\theta(p,r))\,dS(p)d\omega(p,dS(r)) .$$

The differential small surfaces $dS(p)$ and $d(\text{proj}_{\perp(r,p)}(S(p)))$ are related by the relation:

$$d(\text{proj}_{\perp(r,p)}(S(p))) = \cos(\theta(p,r))dS(p) .$$

The quantity $\cos(\theta(p,r))$ is positive. Here, $\theta(p,r)$ is the angle between the inner normal $\overrightarrow{n(p)}$ to the surface at the point p and the direction $\overrightarrow{pr}$. With a view to physical relevance, the meaning of the angle $\theta(p,r)$, hence of the infinitesimal heat flow $dQ^a(p)$, excludes discontinuity at the point p. Thus, we can write the above expression for $dQ^a(p)$ if we merely know the inner normal at the enclosure boundary S. Therefore, we shall assume the enclosure boundary S to be sufficiently smooth so that it possesses a continuously varying inner normal $\overrightarrow{n(p)}$ at any point p. Then, we can express the cosine of $\theta(p,r)$ in the form:

$$\cos(\theta(p,r)) = \frac{\overrightarrow{n(p)} \cdot \overrightarrow{pr}}{\|r-p\|_{R^3}} ,$$

---

[3]We recall that the emissivity is the property of a material that describes its ability to emit radiation as compared with the emission from a blackbody material at the same temperature; whereas the absorptivity is the property of a material that gives the fraction of energy incident on the material that is absorbed.

with $\|r - p\|_{R^3}$ being the Euclidean distance between the points r and p of the physical space R^3. However, enclosures may have regions where this consideration will not apply. This is for example the case at vertices of a rectangular furnace. In such situations it is possible, as will be shown in Chap. 4, to approximate our boundary surface continuously by another surface where our assumption holds.

Remark 1 (Non-convex Enclosures) *The term* $\cos(\theta(p, r))$ *is positive or equal to zero, hence the angle* $\theta(p, r)$ *must range between* 0 *and* $\pi/2$. *For otherwise the point* r *is not seen when looking from* p, *and there exists a geometrical shadow zone between these two points.*

Remark 2 *The following relation clearly holds*

$$\int_{\omega(p,S)} \cos(\theta(p, r))d\omega(p, dS(r)) = \pi$$

if the solid angle subtended by the enclosure surface S *at the point* p *extends to* 2π *steradian; that is,* $\omega(p, S) \equiv 2\pi$ *steradian.*

To convince ourselves we proceed as follow. First we view $\theta(p, r)$ as the zenith angle of the point r in a spherical coordinate system centered at p and let $\varphi(p, r)$ designate the azimuth angle of the point r. Then, we merely observe that the differential solid angle $d\omega(p, dS(r))$ is given as:

$$d\omega(p, dS(r)) = \sin(\theta(p, r))d\theta(p, r)d\varphi(p, r).$$

On assuming that

$$\theta(p, r) \in [\theta_1(p), \theta_2(p)] \subseteq [0, \pi/2]$$

and

$$\varphi(p, r) \in [\varphi_1(p), \varphi_2(p)] \subseteq [0, 2\pi] ,$$

we arrive at the relation:

$$\int_{\omega(p,S)} \cos(\theta(p, r))d\omega(p, dS(r)) = f(p) ,$$

with

$$f(p) = \frac{1}{2}(\varphi_2(p) - \varphi_1(p))(\sin^2(\theta_2(p)) - \sin^2(\theta_1(p))).$$

If the surface at the point p is sufficiently smooth so that the solid angle $\omega(p, S)$ extends over 2π steradian; that is,

$$[\theta_1(p), \theta_2(p)] = [0, \pi/2]$$

and

$$[\varphi_1(p), \varphi_2(p)] = [0, 2\pi] ,$$

we readily obtain:

$$\int_{\omega(p,S)} \cos(\theta(p,r))d\omega(p,dS(r)) = \pi .$$

By Kirchhoff's law, the fraction of blackbody emission radiated by a surface of an opaque body is equal to the fraction of incident radiative energy absorbed by the surface (i.e. $\alpha = \varepsilon$). It comes that:

$$dQ^a(p) = \varepsilon(p)I^i(r,p)\cos(\theta(p,r))\, dS(p)d\omega(p,dS(r)).$$

The surface is assumed to be sufficiently smooth so that it possesses a continuously varying inner normal at any point of it. Integrating the infinitesimal heat flow over all $\omega(p,S)$ incoming directions and dividing the result by the surface area gives the total radiative heat flux absorbed by a unit surface; that is,

$$q^a(p) = -\int_{\omega(p,S)} \varepsilon(p)I^i(r,p)\cos(\theta(p,r))\, d\omega(p,dS(r)).$$

As illustrated in Fig. 2.11, the absorbed energy rate is taken as positive in the direction of the outward surface normal (going into the medium), so that $q^a(p)$ the flux going into the surface at the point p is negative since $\cos(\theta(p,r))$ is positive. Similarly, the radiative energy emitted by the surface, integrated over all outgoing directions, follows as:

$$q^e(p) = \int_{\omega(p,S)} \varepsilon(p)I_b(p)\cos(\theta(p,r))\, d\omega(p,dS(r)).$$

The emitted flux $q^e(p)$ is positive since it is going into the medium.

The net radiative heat flux is calculated by adding both contributions of emitted and absorbed energy. Accordingly,

$$q(p) = \varepsilon(p)\int_{\omega(p,S)} (I_b(p) - I^i(r,p))\cos(\theta(p,r))\, d\omega(p,dS(r)). \tag{2.4}$$

The quantity $q(p)$ represents the net effect of radiative interactions occurring at the surface point p. It defines the rate at which energy would have to be supplied to or removed from the surface by some means other than radiation to make up the net radiative loss and thereby maintain the specified surface temperature at the location p. In industry, the net flux distribution $q(p)$ is one of the important parameters used in assessing the combustion chamber performance.

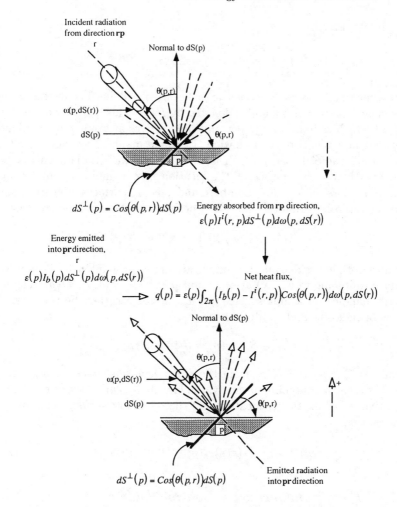

Figure 2.11. Geometry for derivation of energy balance on a unit surface

For the sake of convenience we shall assume in the remaining of this book that the solid angle $\omega(p, S)$ subtended by the enclosure surface S at the point p extends over 2π steradian. The integral of the blackbody intensity at the location p over all 2π incoming directions is denoted by $e_b(p)$. This quantity, also termed blackbody emissive power, represents the rate at which radiative energy is emitted by an elemental blackbody surface. Thus, (2.4) is rewritten as:

$$q(p) + \varepsilon(p) \int_{2\pi} I^i(r, p) \cos(\theta(p, r)) \, d\omega(p, dS(r)) = \varepsilon(p)e_b(p). \qquad (2.5)$$

The integral term

$$\int_{2\pi} I^i(r,p)\cos(\theta(p,r))d\omega(p,dS(r))$$

appearing in (2.5) is known as the area density of radiant flux, or the surface irradiance, at the point p and is denoted by $H(p)$. This quantity defines the rate at which radiant energy per unit area arrives at point p by combined emission and reflection of radiation. In a like manner, we can introduce the surface radiosity $J(p)$ which is the rate at which radiant energy per unit area leaves the surface at the location p by combined emission and reflection of radiation. For Lambert and opaque surfaces, the irradiance and the radiosity are interrelated by the following heat balance relation:

$$J(p) = \varepsilon(p)e_b(p) + (1 - \varepsilon(p))H(p) .$$

Since both emissions and reflections are diffuse, the intensity leaving the surface (outgoing intensity) at any point p is $I^o(p) = (1/\pi) \cdot J(p)$. The net energy balance on a unit surface given by (2.5) can then be rewritten in terms of radiosity and irradiance as:

$$\forall p \in S, \quad q(p) = J(p) - H(p) .$$

Eliminating $J(p)$ and $H(p)$ from the above equation gives a simple relation between the outgoing intensity, the blackbody emissive power and the radiative heat flux at any location p of the boundary surface. The form of this relation is:

$$\varepsilon(p)I^o(p) = \frac{1}{\pi}(\varepsilon(p)e_b(p) - (1 - \varepsilon(p))q(p)) . \tag{2.6}$$

2.6.1 Energy Balance on a Control Volume

At any point p within the enclosure volume, the intensity of incident radiation imparts an infinitesimal heat flow on a control volume in the amount of:

$$dQ_V^a(p) = a(p)I^i(r,p)dV(p)d\omega(p,dS(r)) .$$

Integrating this infinitesimal heat flow over all 4π steradian incoming directions and dividing the result by the measure of the control volume gives the total absorbed radiative heat flux; that is,

$$q_V^a(p) = -\int_{4\pi} a(p)I^i(r,p)d\omega(p,dS(r)) .$$

As shown in Fig. 2.12, the absorbed energy is taken as positive in the direction towards the control volume so that the radiative flux $q_V^a(p)$ going into the volume is negative.

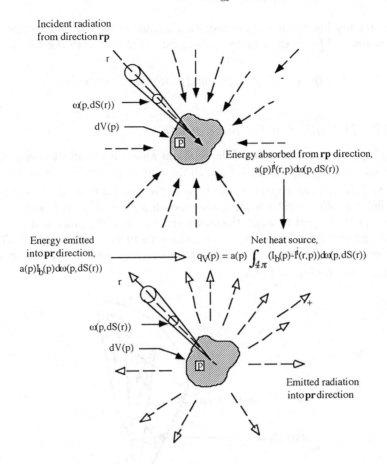

Figure 2.12. Geometry for derivation of energy balance on a control volume

In local thermodynamic equilibrium, it follows from Kirchhoff's law that the fraction of blackbody emission radiated by a control volume is equal to the fraction of incident radiative energy absorbed by the same volume. Therefore, the radiative heat spontaneously emitted from it, and integrated over all 4π outgoing directions follows as:

$$q^e_V(p) = \int_{4\pi} a(p)I_b(p)d\omega(p, dS(r)) \, .$$

The net outflow of radiant energy per unit volume, which is the divergence of a net radiative heat flux vector, also referred to as the radiative heat source and denoted by q_V, is calculated by adding both contributions. Accordingly,

$$q_V(p) = a(p) \int_{4\pi} (I_b(p) - I^i(r, p))d\omega(p, dS(r)). \qquad (2.7)$$

Since the blackbody intensity of radiation is equal to the blackbody emissive power divided by π, the equation above can be rewritten in the form:

$$q_V(p) + a(p) \int_{4\pi} I^i(r,p)d\omega(p, dS(r)) = 4a(p)e_b(p). \qquad (2.8)$$

2.6.2 Definition of the Solid Angle

To complete the formulation of the relations above, it remains to define the differential solid angle $d\omega(p, dS(r))$. To this end we denote by r an arbitrary point of the interior surface S and by $\overrightarrow{n(r)}$ the normal to S constructed at the point r. We consider a piecewise smooth surface $S(r)$, containing the point r, with respect to which the outward direction is the positive direction of the normal. Now, let the point $p \in D$ be such that: at any point $s \in S(r)$ the vector $\overrightarrow{sp}$ forms an acute angle, or at most a right angle with the normal $\overrightarrow{n(s)}$, so that $\cos(\theta(s,p)) \geq 0$.

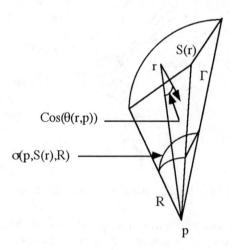

Figure 2.13. Geometry for definition of the solid angle

From the point p we draw radius vectors to all the points of the surface $S(r)$. These radius vectors fill the domain bounded by the surface $S(r)$ and the conical surface Γ, the latter being generated by those radius vectors which terminate at points on the edge of $S(r)$ as illustrated in Fig. 2.13. Taking the point p as center, we draw a sphere of arbitrary radius R. We denoted by $\sigma(p, S(r), R)$ the portion of this sphere which is enclosed in the cone referred to above. Then, the ratio

$$\omega(p, S(r)) = \frac{mes(\sigma(p, S(r), R))}{R^{n-1}},$$

with n being the dimension of the physical space $\mathbb{R}^n$ being considered, is independent of R. It is called the solid angle subtended by the surface $S(r)$ at the point p. The solid angle $\omega(p, S(r))$ is defined to be:

$$\omega(p, S(r)) = -\int_{S(r)} \frac{\partial}{\partial n(r)} \left(\frac{1}{\|r - p\|_{\mathbb{R}^3}} \right) dS(r) .$$

Thus, the differential solid angle is given as:

$$d\omega(p, S(r)) = -\frac{\partial}{\partial n(r)} \left(\frac{1}{\|r - p\|_{\mathbb{R}^3}} \right) dS(r) ,$$

or equivalently,

$$d\omega(p, S(r)) = \frac{\cos(\theta(r, p))}{\|r - p\|_{\mathbb{R}^3}^2} dS(r) .$$

The solid angle is a dimensionless quantity, but in a three dimensional space it is said to be measured in steradian, just as the plane angle in a two dimensional space is said to be measured in radians.

Now we may rewrite (2.5) and (2.8) in the form:

$$q(p) + \varepsilon(p) \int_S I^i(r, p) \frac{\cos(\theta(r, p)) \cos(\theta(p, r))}{\|r - p\|_{\mathbb{R}^3}^2} dS(r) = \varepsilon(p) e_b(p), \qquad (2.9)$$

for $p \in S$, and

$$q_V(p) + a(p) \int_S I^i(r, p) \frac{\cos(\theta(r, p))}{\|r - p\|_{\mathbb{R}^3}^2} dS(r) = 4a(p) e_b(p) \qquad (2.10)$$

for $p \in D - S$.

These equations are the energy balance relations of radiative heat transfer for a unit surface and for a control volume respectively. The absorption coefficient $a(p)$ varies with temperature and medium composition (gas-filled furnaces).

2.6.3 Radiation Pressure and Stress

As mentioned already, radiation occurs when a molecule of matter emits or absorbs a train of electromagnetic waves or photons by lowering or raising its molecular energy levels. Thus, we may thing of thermal radiation as photons traveling in straight lines at the speed of light c and carrying an energy $E = h\nu$, where h is Planck's constant and ν is the frequency of photon energy. These particles should transport momentum in the amount of $M = E/c$.

Therefore, the incident beam of radiation transmits an infinitesimal incident energy flow at any point p of the surface in the amount of:

$$dQ(p) = I^i(r, p) \cos(\theta(p, r)) \, dS(p) d\omega(p, dS(r)).$$

When radiant energy arrives at the enclosure boundary S a momentum transfer takes place. This momentum transfer implies that a stream of photons exerts pressure on the surface S defined as the radiation pressure.

The stress tensor $\sigma(p) = (\sigma_{ij}(p))_{1 \leq i,j \leq 3}$ at the point p due to radiation is defined to be:

$$\sigma_{ij}(p) = -\frac{1}{c} \int_{2\pi} I^i(r,p) n_i(r,p) n_j(r,p) d\omega(p, dS(r)) ,$$

where $n_i(r,p)$ is the direction cosine of the direction of travel of radiation with respect to the $i - th$ axis.

Hence, the radiation pressure at the point p is given by:

$$P(p) = -(\sigma_{11}(p) + \sigma_{22}(p) + \sigma_{33}(p)) ;$$

accordingly,

$$P(p) = \frac{1}{c} \int_{2\pi} I^i(r,p) \cos^2(\theta(p,r)) d\omega(p, dS(r)).$$

Using the relation between $d\omega(p, dS(r))$ and $dS(r)$, we may rewrite the matrix elements of radiation stress tensor and the radiation pressure respectively in the form:

$$\sigma_{ij}(p) = -\frac{1}{c} \int_S I^i(r,p) n_i(r,p) n_j(r,p) \frac{\cos(\theta(r,p))}{\|r - p\|^2_{\boldsymbol{R}^3}} dS(r) ,$$

and consequently,

$$P(p) = \frac{1}{c} \int_S I^i(r,p) \frac{\cos(\theta(r,p)) \cos^2(\theta(p,r))}{\|r - p\|^2_{\boldsymbol{R}^3}} dS(r).$$

The above radiation stress tensor and pressure, while measurable, may not be negligible for certain engineering radiative heat transfer applications involving momentum considerations.

We now pass on to a brief review of existing numerical methods of solution which the reader has probably already met in the literature.

3

Some Computational Methods

Before entering into the analysis of the radiative heat transfer equation, it is necessary to become familiar with existing techniques for prediction of radiant heat exchanges in enclosures. This chapter begins with a review of the computational techniques presently used for calculating the net radiative heat flux. Further on, the radiative heat transfer equation studied in this book is described.

Due to the complexity inherent in the physical and mathematical descriptions of radiative heat transfer in industrial furnaces and combustion chambers, various approximate techniques for solving radiative heat exchange problems have been developed through the years. Basically, we can divide these methods into two main groups:
- Methods based on the radiative transfer equation;
- Methods based on the net energy balance relations.

3.1 Directional Equation Methods

The intensity of incident radiation is the most important function of this group of methods. Due to its direction dependency, there exists an infinite number of independent values of intensity at a given point of the medium. This is the main difficulty to be overcome, since one cannot handle an infinite number of directions. In the integrated directional equation methods, only a finite number of directions are considered. The techniques that belong to this group are the flux, the discrete ordinates, the moments, the spherical harmonics and the discrete transfer methods.

3.1.1   The Flux and the Discrete Ordinates Methods

The flux methods are all derived from the work of Schuster [50]. They were originally devised for solution of astrophysical problems under the assumption that heat transfer by radiation is important in only one predominant direction.

The flux methods are based on a discrete representation of the directional variation of radiative intensity. In this group of method, a numerical solution of the radiation problem is obtained by solving the integrated directional equation of radiative transfer for a set of discrete directions spanning the total solid angle range 4π steradian. As such the methods are simply a finite representation of the directional dependence of the radiative heat transfer equation.

When the solid angle about a given location is divided into two directions within which the directional dependence of the intensity is assumed, the method is known as the two-flux method (Schuster–Schwarzschild, Schuster–Hamaker) [2, 51].

Here, the integral of the intensities over their corresponding solid angle can be interpreted as a one way radiative heat flux and is treated as an unknown. Then, the integrated directional equation of radiative heat transfer is written for the chosen directions.

This produces a first order differential equation for the one-way flux whose direction coincides with that previously chosen. For each direction, two one way fluxes are sought. Thus, the number of equations obtained is twice the number of chosen directions.

Each pair of first order flux equations corresponding to a given direction can be combined to yield a second order differential equation if required. These new equations were found to be compatible with second order equations of mass and energy transfer.

It was realized in the 1970s that the flux methods offered a relatively easy way of incorporating radiation calculations into computational fluid dynamic computer codes. Practically throughout this period, the work on development and validation of the flux models progressed at Imperial College [52, 53], the University of Sheffield [51], and the IFRF [54, 55].

When the solid angles are divided into more than two directions within which the directional dependence of the intensity is assumed, the method is known as the multi-flux, discrete ordinates, or S_N-method. Discrete Ordinates methods do not demand any assumption concerning the angular distribution of intensity. In this method, the working equations are written in a finite difference form and total integration over the solid angle is performed using numerical quadratures [56]-[58].

3.1.2   The Milne–Eddington or Moment Method

In the flux method, the directional dependence of the intensity is an arbitrary function without any adjustable parameters. More accurate approaches use polynomial functions to describe the intensity within the solid angle. The variables of these polynomials are the directional cosines, and their coefficients are made positionally dependent. Substitution of this prescribed distribution of the intensity into the integrated directional equation of radiative heat transfer yields a set of differential equations for the coefficients of the polynomials. Elimination of these coefficients gives a set of second order differential equations for the radiative heat fluxes. Examples on the moment method can be found in the literature [49],[59]-[61].

3.1.3   The Spherical Harmonics or P_N-Method

The spherical harmonics methods [62, 63] are based on the method of separation of variables and on the generalized Fourier series. In these methods, the intensity is expanded into spherical harmonics series with truncation to N terms. Usually, odd expansions $N = 1$, $N = 3$, etc. are used. This is for two reasons [12]. Firstly, the even terms are difficult to incorporate into useful engineering boundary conditions. The odd-order expansions work well in the so-called Marshak's boundary conditions, which have a physical interpretation in terms of the net radiative heat flux for first order terms. Secondly, for one and two dimensional problems, it has been found that second order terms are either negligible or can be eliminated of the formulation of the radiative transfer equation or its moments. Thus, second order expansions often provide only marginal increases in accuracy over the first order solutions. Substitution of the expanded series into the moments of the differential form of the integrated directional equation of radiative transfer yields a set of ordinary differential equations.

3.1.4   The Discrete Transfer Method

In the discrete transfer method [64], the medium and its boundary are subdivided into isothermal volumes and surface elements having constant material properties. The integrated directional equation of transfer is solved iteratively along an arbitrary chosen ray path. The ray is traced along its way from a given point untill the first boundary is encountered. Throughout each direction, the intersection of the ray and the boundary is used as origin point r into (2.3) and the radiative intensity at this starting point is assumed. Solving the transfer equation gives the value of the intensity at the panel central point for a ray coming from the chosen direction. Integration over the solid angle by means of quadratures results in the net incoming radiative flux. Calculation of the heat sources in volume cells is accomplished by a control volume approach tracing each ray and determining the amount of energy it gains or looses when going through a given cell. Developments on the discrete transfer method are described in the literature [64]-[67].

3.1.5 Concluding Remarks

A comprehensive review of the above mentioned numerical methods can be found in the radiation literature [49],[68]-[71]. In heat radiation modeling, these methods often use mean values of the intensity obtained upon integration over an appropriate range of the solid angle. A characteristic feature of this group of methods is that the working equations are differential ones and they can be easily linked with the enthalpy and mass transport equations. The methods suffer from so called ray effect. Since radiant energy is allowed to stream only along some discrete directions, radiation from an isolated source may remain unseen by a point unless the point lies along an ordinate direction. Thus, the above mentioned methods cannot deal with the far distance interactions that characterize radiative heat exchanges. A good treatment of this feature can be achieved only upon using integral equations. A measure of radiation interaction between two points r and p, or the ability of a given path (r, p) within the enclosure to attenuate radiation, is the optical depth. The optical depth is the integral of the absorption coefficient along a line of sight. It is written as:

$$\kappa(r, p) = \int_{(r,p)} a(u) dL(u) .$$

3.2 Net Energy Balance Methods

The net-radiation method simplifies the problem of radiation heat exchange in enclosures by casting the formulation in terms of the net radiative energy at a given surface. It was first devised by Hottel [3, 4] and latter developed in a different manner by Poljak [72]. The net-radiation methods are based on the energy balance relations of radiative heat transfer. The techniques that belong to this group are the probabilistic and the finite element methods.

3.2.1 Probabilistic Methods

The probabilistic methods are variants of the Monte Carlo technique. The Monte Carlo technique is a method of statistical sampling of stochastic physical events, used to estimate the average behavior of a given system.

The methods have arisen in those applications where the basic mathematical problem itself consists of the investigation of some random process.

It was first devised for solving problems in neutron physics which where posed in a probabilistic form [73]. In this model, finite numbers of energy bundles are followed through their transport histories. The radiative behavior of the system is then estimated from the average behavior of these bundles. Theoretically, the energy bundle is randomly generated at a point belonging either to the boundary or to the furnace interior. The wavelength of this bundle, as well as its direction, are also obtained by random generation. The direction of this bundle depends on the directional characteristic of the material at the point where the energy portion originated. The energy portion is traced on its way through the participating medium where it can be attenuated, the probability of this event being a function of the enclosure radiative properties.

We should stress the fact that this model neither provides us with the solution to the radiative behavior of the system, nor gives us indications of the existence of a solution. Besides, the working of the Monte Carlo methods involve the task of connecting a computer to the physical process of radiation with a random feature attached to it. This, however, is not done in practice because of the large computer storage required.

Instead, pseudo-random[1] sequences of numbers are generated and used as in the number theory. An estimation of the error included in this process, which is often expressed as the standard deviation, has the poor feature from the stand point of numerical analysis that the numbers cannot be duplicated easily. Thus, the result of a computation cannot be readily checked.

For general questions concerning the Monte Carlo technique and a review of literature dealing with the probabilistic methods we refer the reader to radiative heat transfer monographs and text books [2, 49],[73]-[76].

3.2.2   Finite Element Methods

The finite element method utilizes the (discretized) integral forms of energy balance equations. There are several variations of this approach. Most of them require a subdivision of both the volume and the surface into a finite number of subdomains (finite elements). Within these elements the variation of the unknown functions and the radiative properties are assumed. If each element is taken to be isothermal, then the method is known as a zoning technique [77, 78].

The finite element methods lead to a set of simultaneous algebraic equations for the unknown heat fluxes. The methods use large numerical grids and their computer running times tend to be long. Computational economy can be achieved by using the product integration method or by transforming the volume integrals arising in the net balance energy relations into

---

[1]Here is meant a deterministic sequence of numbers defined mathematically.

surface integrals. This latter version of the finite element method is known as the boundary element method [45],[79]-[86].

A considerable and comprehensive review on the net radiation methods can also be found in the radiative heat transfer literature [49],[68]-[71], [87]. All these methods lead to sets of algebraic equations linking radiative emissive powers, radiative heat fluxes and radiative heat sources. The methods can deal with the far distance interactions that characterize the radiative heat exchange. The main drawback of these techniques is their excessive computer running time. This is due to the fact that the numerical grid at the enclosure boundary is often an extension at the boundary of the grid used in partitioning the enclosure volume.

3.3   Concluding Remarks

The methods outlined in the sections above were developed to solve the radiative heat transfer equation. They all give, numerically, candidates to the solution of radiation heat transfer problem in enclosures. But how reasonable and consistent, and reliable these approximations are; that is, with which accuracy they approximate the actual solution, remain unanswered. Each of these methods was applied to simple one dimensional gray gas problems. When they were extended to multidimensional geometries with nongray medium and with various boundary conditions, drawbacks emerged in every case [12, 13]. The drawbacks were related either to establishing the solvability of the mathematical problem (existence and uniqueness of the solution) or to obtaining an efficient computational algorithm. Heat transfer by radiation still demands a search for a novel numerical algorithm based rather on theoretical grounds than on intuition.

In the radiative heat transfer community, preocupation with combustion chambers reality leads to resistance and rejection of methods based on advanced mathematical tools which do not parallel with this reality. Here the notion of reliability is controversial since tolerance to deviations from a rigorous formulation varies widely. There are researchers who are perfectly content with existing methods. They have no qualms when these methods exhibit large differences, up to 200-300%, in their predictions as pointed out in an article by Tong and Skocypec [13], or when positive and negative values occur in the predictions of incident radiative heat flux distributions at the enclosure boundary. But they are hostile to the use of advanced mathematics to address the radiant energy exchange problem, levelling the charge unphysical. In short, all mathematical developments and conclusions are subjected to the test of immediate physical relevance and their numerical results are compared against existing "solutions".

Finding another numerical method that will attempt to solve the problem of radiation heat transfer by comparing its numerical results with those obtained using the methods outlined in this chapter is not hard. Into this classical engineering problem we shall not enter. However, providing a fundamental, efficient, and well-constructed algorithm whose accuracy can be assessed by using stringent mathematical rules without obscuring the essential nature of radiation is much more challenging.

It is certainly important to develop a good mathematical formulation of the radiative transfer problem and adequate treatments for the fundamental equations. Presently, for example, there is no technological device or instrument which can be used to provide the net radiative heat source measurements inside the enclosure volume. One must rely on numerical predictions. Therefore, it is also essential to establish that this formulation actually does furnish efficient and reliable computational solutions. The new formulation should possess the advantages of the previously described methods and should overcome their disadvantages. Here, we wish to draw attention to the fundamental fact that the meaning of a numerical solution (or numerical method), specifically for radiative heat transfer phenomena which is governed by an integro-differential equation, is not precise unless it is supplemented by an estimate of the errors occurring; that is, unless it is accompanied by definite knowledge of the degree of accuracy attained.

3.4   The Boundary Value Equation

We pause briefly in this section and summarize the previously derived radiation equations we wish to study.

When scattering can be neglected, the integrated directional equation of radiative heat transfer is given by (2.3). To calculate the intensity $I^i(r,p)$ of incident radiation at a moving point p, the intensity $I^o(r)$ originating from point $r \in S$ should be known. The outgoing intensity $I^o(r)$ also depends on the intensities at other points of the enclosing surface. This dependence makes the integrated directional equation of radiative transfer complex.

An alternative form of the equation governing radiative heat transfer is the boundary value equation. This formulation is an equivalent integral equation formulated on the boundary of the domain considered. For radiative heat transfer in enclosures, the boundary value equation results from the radiative energy balance relations. The outgoing intensity at any point $r \in S$ is first determined in terms of blackbody emissive power and net radiative heat flux using (2.6). The resulting intensity of incident radiation at the point $p \in D$, provided that the emissivity $\varepsilon(r)$ is not equal to zero at the origin r of the line of sight, has the form:

$$I^i(r,p) = \frac{1}{\pi}\left(e_b(r) - \frac{1-\varepsilon(r)}{\varepsilon(r)}q(r)\right)\tau(r,p) + \frac{1}{\pi}L(r,p) \; ,$$

with

$$L(r,p) = \int_{(r,p)} a(u)e_b(u)\tau(u,p)dL(u) \ .$$

Substitution of the above relations into the radiative heat balance equations on a unit surface (2.9) and on a control volume (2.10) yields the boundary value equation for heat transfer by radiation.

Let us define the kernel $K(r,p)$ by the relation:

$$\forall r \neq p \in S, \quad K(r,p) = \frac{\cos(\theta(r,p))\cos(\theta(p,r))}{\pi \, \|r - p\|_{\boldsymbol{R}^3}^2} \ .$$

Then, the boundary value equation of radiative transfer on a unit surface reads:

For $\varepsilon(p) = 0$ almost everywhere[2],

$$q(p) = 0.$$

For $\varepsilon(p) = 1$ almost everywhere,

$$q(p) = e_b(p) - \int_S (e_b(r)\tau(r,p) + L(r,p))\, K(r,p)dS(r) \ .$$

For $0 < \varepsilon(p) < 1$,

$$\frac{q(p)}{\varepsilon(p)} + \int_S \left(\left(e_b(r) - \rho(r)\frac{q(r)}{\varepsilon(r)}\right)\tau(r,p) + L(r,p)\right) K(r,p)dS(r) = e_b(p),$$

with $\rho(r) = 1 - \varepsilon(r)$ being the reflectivity at the point r.

The heat source distribution inside the medium satisfies the relation

$$q_V(p) = a(p)\left(4e_b(p) - \int_S (J(r)\tau(r,p) + L(r,p))\, K_V(r,p)dS(r)\right),$$

with the kernel $K_V(r,p)$ defined as:

$$\forall r \in S, \ \forall p \in D - S, \quad K_V(r,p) = \frac{\cos(\theta(r,p))}{\pi \, \|r - p\|_{\boldsymbol{R}^3}^2} \ .$$

The radiation pressure at any boundary point $p \in S$ is given by:

$$P(p) = \frac{1}{c} \int_S (J(r)\tau(r,p) + L(r,p))\, K_o(r,p)dS(r) \ ,$$

---

[2]Everywhere except on a neglected set of points; that is, a set of points either countable and finite or a Borel set of measure zero. We say that a set E of points is of measure zero if for any $\eta > 0$ there exists a countable number of spheres, each with measure less than η, covering E.

with the kernel $K_o(r, p)$ defined by the relation:

$$\forall r \neq p \in S, \quad K_o(r, p) = \frac{\cos(\theta(r, p)) \cos^2(\theta(p, r))}{\pi \, \|r - p\|_{\mathbf{R}^3}^2} \, .$$

Remark 3 *For non-convex enclosures, we need not use the shadow zone function $\chi(r, p)$ defined in Sect. 2.4. Instead, because of Remark 1, we can redefine the kernels above as follows:*

$$
\begin{aligned}
K(r, p) &= \frac{\cos(\theta(r, p)) \cos(\theta(p, r))}{\pi \, \|r - p\|_{\mathbf{R}^3}^2} \quad && \textit{if } 0 \leq \theta(r, p), \theta(p, r) < \frac{\pi}{2}, \\
&= 0 && \textit{otherwise.}
\end{aligned}
$$

$$
\begin{aligned}
K_V(r, p) &= \frac{\cos(\theta(r, p))}{\pi \, \|r - p\|_{\mathbf{R}^3}^2} \quad && \textit{if } 0 \leq \theta(r, p) < \frac{\pi}{2}, \\
&= 0 && \textit{otherwise.}
\end{aligned}
$$

$$
\begin{aligned}
K_o(r, p) &= \frac{\cos(\theta(r, p)) \cos^2(\theta(p, r))}{\pi \, \|r - p\|_{\mathbf{R}^3}^2} \quad && \textit{if } 0 \leq \theta(r, p), \theta(p, r) < \frac{\pi}{2}, \\
&= 0 && \textit{otherwise.}
\end{aligned}
$$

It is noteworthy that to determine the radiative intensity, the radiative heat source and the radiation pressure, the radiative heat flux $q(p)$ must be known. Therefore, we have only to investigate the boundary-value integral equation on the enclosure surface S.

The advantage of working with the integral equation based on the net radiative heat flux are threefold: the domain of dependency is reduced from three to two dimensions in space; the integral operator appearing in the equation is bounded and completely continuous, whereas the differential operator is unbounded; the existing theory of variational principle and subsequent approximation schemes can then be directly utilized whilst solving the integral equation. Once the net radiative heat flux $q(p)$ is determined, the desired intensity of radiation and the heat flux inside the medium are readily calculated.

None of the results presented here is changed if, instead of three dimensional enclosures, we consider furnace geometries in the two dimensional world $\mathbb{R}^2$. For two dimensional convex geometries, the kernels $K(r, p)$, $K_V(r, p)$, and $K_o(r, p)$ are defined to be:

$$K(r, p) = \frac{\cos(\theta(r, p)) \cos(\theta(p, r))}{2 \, \|r - p\|_{\mathbf{R}^2}} \, ,$$

$$K_V(r,p) = \frac{\cos(\theta(r,p))}{2\,\|r-p\|_{\boldsymbol{R}^2}}\,,$$

$$K_o(r,p) = \frac{\cos(\theta(r,p))\cos^2(\theta(p,r))}{2\,\|r-p\|_{\boldsymbol{R}^2}}\,.$$

As observed earlier the meaning of the angles $\theta(p,r)$ and $\theta(r,p)$, hence of the kernels $K(r,p)$, $K_V(r,p)$, and $K_o(r,p)$ excludes discontinuity at the points r and p of the enclosure boundary S. We shall confine our attention to convex enclosures having sufficiently smooth interior surface S so that it possesses a continuously varying inner normal.

We have formulated the mathematical relations describing radiative transfer and reviewed the computational methods presently used for thermal radiation applications. For further details of the methods described, the reader is referred to textbooks on heat transfer by radiation [2, 49, 68].

Returning on to the purpose of the present treatise, we will continue our development with three dimensional furnaces. We shall assume in the following chapters that the emissivity of the furnace interior surface, at any point, ranges in the interval $]0,1[$ and, hence there exists a number $\varepsilon_o > 0$ such that:

$$\forall p \in S, \quad \varepsilon(p) \geq \varepsilon_o > 0\,.$$

4

Mathematical Model

In this chapter we shall examine the boundary value equation of radiative transfer in enclosures and develop the fundamentals for its solution. We shall begin with the analysis of the boundary value equation of radiative heat transfer in absorbing-emitting media. Our purpose is to formulate a structural problem and to show that it leads to a uniquely determined solution. We shall investigate the solvability of the problem we have gone into and, in particular, characterize it by further conditions which may be imposed in addition to the equation itself. Although we omit explicitly the wavelength variable λ, the developments given here and in the next chapter remain valid on a strictly monochromatic basis and nothing need be changed in our notations and proofs.

4.1 Subsidiary Conditions

We can classify the problems to which radiative heat transfer theory is applied in practice into two categories: the determination of radiant energy distribution throughout a medium in which the temperature distribution, the species concentrations, the absorption, emission, and scattering properties are given; the determination of temperature distribution and radiating energy distribution in a given medium.

The first category, while requiring some preliminary experimental and physical considerations, can be considered as a mathematical problem since it involves the solution of a complex integro-differential equation.

The second category requires a simultaneous solution of the fluid flow, the energy and the mass conservation equations.

Solving problems within both categories require knowledge of the absorption and emission properties of the medium. The absorption and emission coefficients can be prescribed constant values trough the medium, or they can be expressed locally as functions of the state variables [49, 68, 69, 88, 89].

We wish to determine the net heat flux distribution $q(p)$ over the enclosure boundary S and it is desirable to derive a general mathematical procedure applicable to any enclosure filled in with an absorbing-emitting medium. Therefore, the subsidiary conditions required for the radiative heat exchange calculations will consist of a prescribed temperature distribution within the enclosure volume and over its interior surface. We will also assume that the emissivity values at the boundary surface are known. There are, however, no a priori conditions for the behavior of the sought for function $q(p)$ at any part of the enclosure boundary S.

4.2   Analysis of the Boundary Value Equation

The main objective in solving the boundary value equation is to determine the net radiative heat flux $q(p)$ over the enclosure boundary S. The purpose of this section is to clarify the extremely important question concerning the class of admissible functions for the net heat flux. It is important to define in a clear and precise way the class of functions to be considered as well as the appropriate norms, in order to derive correct convergence properties and error estimates. This class has to be broad enough for a functional to achieve its supremum (least upper bound) and/or its infinimum (greatest lower bound). It will serve as the conceptual substance from which our formulation of physical reality is fashioned.

To approach this problem, we foremost extend the set of functions to include all functions which are integrable in the Lebesgue sense. Here, we view the Lebesgue integral as an extension of the Riemann integral in the sense that every Riemann-integrable function is also Lebesgue-integrable to the same value. We recall that there exist some functions, such as the Dirichlet function which takes the value 1 on rational numbers and 0 on irrational numbers, which fail to Riemann integrability but are Lebesgue integrable.

From the calculation view point, and particularly for our applications, the kind of functions that are Lebesgue integrable but not Riemann integrable do not occur. Thus, it is not for computational reasons that we use the notion of Lebesgue integral. The Riemann integral of a continuous function and of a function that is continuous except at a finite number of points is exactly the same as the Lebesgue integral of that function. In this book, the Lebesgue integral provides a structural and consistent unity to our developments. From a philosophical view point, the relationship of Lebesgue to Riemann integrable functions is similar to that of real to rational numbers. Concrete calculations require only rational numbers, but mathematics needs irrational numbers. The totality of real numbers, rational plus irrational, has an inner consistency absent from the class of rational numbers alone. Similarly, for most concrete calculations the notion of Riemann integral is adequate, but theorems which we shall use are more easily formulated and proved within the class of Lebesgue-integrable functions.

In the remaining of this treatise we shall employ the element of Lebesgue measure $\mu(dS(p)) \equiv dS(p)$, where μ represents the positive Lebesgue measure, in the integrals.

The formulation of our boundary value equation is based on the tacit assumption of sufficiently smooth boundary S which possesses a continuously varying inner normal. Presently, we merely assume:

The interior surface S possesses continuous curvatures; i.e. that it can be covered by a finite number of spheres which have the property that, singling one spatial coordinate the part of the surface contained in each of these spheres is congruent to a surface represented by a function $z = \psi(x,y)$, where ψ is continuous[1] and has continuous derivatives up to and including the second order.

To see if there is any real substance in this hypothesis we shall aim first to establish its plausibility.

Let S be a surface on which the above consideration is not satisfied. To justify our assumption we must use a sufficiently strong concept of approximation of a surface S by another surface S_η satisfying the above requirement, with $\eta > 0$ being a size parameter.

Whenever continuous curvatures occur at the boundary S, it is enough to require that the surface S_η approximates S pointwise and the inner normal to the surface S_η approximate those of S. Analytically we may define approximation of the surface S by the surface S_η in the stronger sense as follows:

Let S be transformed pointwise into S_η by relations of the form:

$$x_\eta = x + f(x,y,z), \quad y_\eta = y + g(x,y,z), \quad z_\eta = z + h(x,y,z);$$

with f, g, h being functions continuous that possess continuous derivatives up to and including the second order throughout the boundary S, and the

---

[1] The following definition should be recalled: a function is said to be continuous if sufficiently small changes in its arguments produce arbitrarily small changes in the function values.

functions f, g, h and their derivatives are less in absolute value than a small number $\eta > 0$. Then, we say that the surface S is approximated by the surface S_η with the degree of accuracy η. When η approaches zero we say that S is deformed continuously into S_η.

With these considerations it is not difficult to see that the radiative heat flux at the surface S_η, for corresponding temperatures distributions and material properties, approximates the radiative heat flux at the surface S arbitrary closely, if the domain S_η differs arbitrarily little from S.

We now insert a discussion of a concept which has reference only to the behavior of a curve in the neighborhood of a point, namely, the curvature.

Remark 4 *If we think of the curve $y = f(x)$ as described uniformly in the positive sense, in such a way that equal lengths of arcs are passed over in equal periods then, the direction of the curve will vary at a definite rate, which we take as a measure of the curvature of the curve. If, therefore, we denote the angle between the positive direction of the tangent and the positive x-axis by α, and if we think of α as a function of the length of arc $s = l(p)$, we shall define the curvature $c(p)$ at the point p corresponding to the length of arc s by the equation*

$$c(p) = \frac{d\alpha}{dl(p)} .$$

We know that $\alpha = \arctan(y')$, and hence by the chain rule the curvature is given by the expression:

$$c(p) = \frac{d\alpha}{ds} = \frac{y''}{(1+y'^2)^{3/2}} .$$

The treatment of the boundary value equation for the net radiative heat flux $q(p)$ is based on certain properties of the positive-valued function $K(r,p)$.

Now, without obscuring the physical nature of radiation phenomena, we proceed to examine the essential features of the kernel $K(r,p)$. The kernel $K(r,p)$ is a function of the variables r and p in the physical space represented by the enclosure boundary $S \subset \mathbb{R}^3$. It is written in the form:

$$K(r,p) = \frac{\cos(\theta(r,p))\cos(\theta(p,r))}{\pi \|r-p\|_{\mathbf{R}^3}^2} ,$$

with $\cos(\theta(r,p))\cos(\theta(p,r))$ being a bounded function which is continuous and defined everywhere over $S \times S$. We observe:

$K(r,p)$ is defined and continuous for all $r \neq p \in S$ and it converges as r approaches p along the space curve $\Gamma(r,p)$, given by the intersection between

the boundary S and the plane perpendicular to the tension plane through p
and containing the vectors $\overrightarrow{rp}$ and $\overrightarrow{n(p)}$, to the finite value $(1/4\pi)(c(p))^2$,
where $c(p)$ is the curvature at the point p along that curve.

The justification proceeds as follows. First we consider the length of arc
$l(u)$ reckoned from a fixed point on the curve $\Gamma(r,p)$ to the current point
u. Then, we may write the term $\cos(\theta(r,p))$ in the form:

$$\cos(\theta(r,p)) = \frac{d\theta(p,r)}{dl(r)} \cdot \|r - p\|_{\boldsymbol{R}^3} .$$

As illustrated in Fig. 4.1, it is not difficult to see that $d\theta(p,r)$ is the angle
formed by the two infinitely closed radius vectors drawn from the point p
to the extremities of the arc $dl(r)$. It is sufficient to stipulate merely that
our kernel converges as r approaches p if the term

$$\frac{d\theta(p,r)}{dl(r)}$$

has a finite value.

For, if every point $u \in \Gamma(r,p)$ is represented in the parametric form
$u(\xi(l(u)), \eta(l(u)))$ with the functions ξ and η being twice continuously

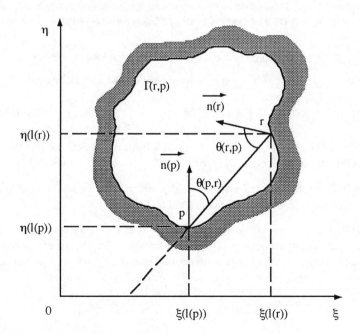

Figure 4.1. Geometry for definition of the kernel $K(r,p)$

differentiable we readily obtain the relation

$$\theta(p,r) = \arctan\left(\frac{\xi(l(r)) - \xi(l(p))}{\eta(l(r)) - \eta(l(p))}\right) ,$$

whence

$$\frac{\cos(\theta(r,p))}{\|r - p\|_{\boldsymbol{R}^3}} = \frac{d}{dl(r)}\left(\arctan\left(\frac{\xi(l(r)) - \xi(l(p))}{\eta(l(r)) - \eta(l(p))}\right)\right) .$$

Applying l'Hospital's rule, upon approaching the fixed point p, we obtain

$$\lim_{r \to p} \frac{\cos(\theta(r,p))}{\|r - p\|_{\boldsymbol{R}^3}} = \frac{1}{2}c(p) .$$

By symmetry it comes immediately that the kernel $K(r,p)$, which may be written in the form

$$K(r,p) = \frac{1}{\pi}\frac{d\theta(r,p)}{dl(p)}\frac{d\theta(p,r)}{dl(r)} ,$$

converges to the finite value $(1/4\pi)(c(p))^2$ for r sufficiently near p, which was to be justified.

It can be seen further that:

$$\forall r, p \in S, \quad K(r,p) \le \max\left\{(1/4\pi)\cdot C_{\max}^2, \sup_{u \neq v} K(u,v)\right\} < +\infty ,$$

with $C_{\max}$ being the maximal curvature of the enclosure surface S.

We ascertain from the considerations above that the Lebesgue integrals

$$\int_S |K(r,p)|\,dS(r) , \quad \int_S |K(r,p)|\,dS(p) , \quad \int_S\int_S |K(r,p)|\,dS(p)dS(r)$$

exist. From these facts we obtain at once a number of general remarks and properties about the kernel $K(r,p)$.

Remark 5 *In a spherical enclosure of finite radius R, the kernel $K(r,p)$ is constant and equal to the limiting value $1/(4\pi R^2)$.*

Remark 6 *For radiative heat transfer, the entire energy leaving any point p of the enclosure boundary S must be incident on the whole surface S. Thus, all fractions of energy leaving any point $p \in S$ and reaching the enclosure boundary must total to unity. Furthermore we have:*

$$\int_S\int_S K(r,p)dS(p)dS(r) = mes(S) < +\infty .$$

This relation is also known as the summation relation for configuration or shape factors.

Remark 7 *The radiation of heat from an enclosure filled in with a homogeneous medium, in the absence of temperature gradients, is characterized at the boundary S by a condition of the form:*

$$\forall p \in S, \quad q(p) = 0 \,.$$

This condition states that the rate of change of energy at the boundary S (with continuous curvatures) must become zero; that is, every point p of the surface S emits precisely as much energy as it absorbs. This state of affairs is an immediate consequence of the Second Law of Thermodynamic, and is called the "no flux condition" in radiative heat transfer. Its analog in mechanics is known as the rigid body movement. From this observation, we obtain by verification that the relation

$$\int_S K(r,p)dS(r) = 1 \,,$$

must be satisfied at every boundary point p. Physical reasoning leads us to the deduction that at any point p of a sufficiently smooth enclosure boundary S, the integral

$$k(p) = \int_S K(r,p)dS(r)$$

which does not depend on the behavior of the medium filling up the enclosure volume is identically equal to unity. Furthermore, the inequality

$$\int_S K(r,p)dS(r) \le 1 \quad p \in S \,,$$

holds. Since, otherwise there will exist a point $p_o \in S$ such that the total sum of fractions of energy leaving the point $p_o \in S$ and reaching the enclosure surface is greater than unity, which contradicts Remark 6.

To evaluate mathematically the integral $k(p)$ defined above, we note first that the differential solid angle subtended at the point p by a small surface $S(r)$ surrounding the point r is defined to be:

$$d\omega(p, S(r)) = \frac{\cos(\theta(r,p))}{\|r - p\|_{\mathbf{R}^3}^2} dS(r) \,.$$

Therefore,

$$\int_S K(r,p)dS(r) = \frac{1}{\pi} \int_{\omega(p,S)} \cos(\theta(p,r))d\omega(p, S(r)) \,,$$

with $\omega(p, S)$ being the solid angle subtended at the point p by the boundary S. As in Remark 2, we view $\theta(p,r)$ as the zenith angle of the point r in a spherical coordinate system with center at the point p and we designate by

$\varphi(p,r)$ the azimuth angle of the point r. Then, the differential solid angle $d\omega(p, S(r))$ is also given as:

$$dw(p, dS(r)) = \sin(\theta(p,r))d\theta(p,r)d\varphi(p,r) .$$

On assuming that

$$\theta(p,r) \in [\theta_1(p),\ \theta_2(p)] \subseteq [0,\ \pi/2] ,$$

and

$$\varphi(r,p) \in [\varphi_1(p),\ \varphi_2(p)] \subseteq [0,\ 2\pi] ,$$

we arrive at the relation:

$$\int_S K(r,p)dS(r) = \frac{(\varphi_2(p) - \varphi_1(p))}{2\pi}(\sin^2(\theta_2(p)) - \sin^2(\theta_1(p))).$$

If the surface at the point p is sufficiently smooth so that the solid angle $w(p, S)$ extends over 2π steradian; that is,

$$[\theta_1(p),\ \theta_2(p)] = [0,\ \pi/2] ,\quad and \quad [\varphi_1(p),\ \varphi_2(p)] = [0,\ 2\pi]$$

we obtain:

$$\int_S K(r,p)dS(r) = 1 .$$

If the surface at the point p is not sufficiently smooth, hence the solid angle does not extend over 2π steradian, we have:

$$\int_S K(r,p)dS(r) \le 1 .$$

The integral $k(p)$ can also be evaluated by application of Stokes' formula which transforms a surface integral taken over an oriented surface S into a line integral taken around the correspondingly oriented boundary of the surface [87, 90].

Remark 8 *From the view point of potential theory, the kernel $K(r,p)$ may be interpreted as product of potential of single dipoles over a curve $\Gamma(r,p)$ in a two dimensional space. The potential of a single dipole, at the fixed point p of a unit mass concentrated at the point r, is defined to be:*

$$\frac{\partial}{\partial n(r)} \log(\|r - p\|_{\mathbf{R}^2}) = \frac{\cos(\theta(r,p))}{\|r - p\|_{\mathbf{R}^2}} .$$

If we fix the point p on the interior surface S, we may consider a mass distribution concentrated on S with the density $(1/\pi) \cdot \cos(\theta(p,r))$ since the inequality $0 \le \cos(\theta(p,r)) \le 1$ holds. Then, the integral

$$\int_S \frac{\cos(\theta(p,r))}{\pi} \frac{\partial}{\partial n(r)}(\|r - p\|_{\mathbf{R}^2})dS(r) = \int_S K(r,p)dS(r)$$

exists and it represents the potential of a double layer with the density $(1/\pi) \cdot \cos(\theta(p,r))$ over the interior surface S.

We desist from further discussions and merely state from the above remarks that the positive-valued kernel $K(r, p)$ is Lebesgue integrable. We also infer from the foregoing discussion that the following integrals

$$\int_S |K(r,p)|^2 \, dS(r) \, , \quad \int_S |K(r,p)|^2 \, dS(p) \, , \quad \int_S \int_S |K(r,p)|^2 \, dS(p) dS(r)$$

remain below a fixed finite bound. This shows that $K(r, p)$ is a square integrable function defined over the domain $S \times S - \{r = p\}$.

Remark 9 *The integral of the function $|K(r,p)|^2$ over the fundamental domain $S \times S$ does not exceed unity. Indeed: let p be a point on the enclosure surface S. We surround the point p with a sphere $B(p, \delta)$ of radius $\delta > 0$ and form the integral*

$$\int_{S \backslash B(p,\delta)} |K(r,p)|^2 \, dS(r) \, .$$

Since $K(r,p) \geq 0$ is a continuous function for all $r \neq p$ and bounded at $r = p$, there follows, from the mean value theorem and Remark 7, the existence of a point $r_o \in S \backslash B(p, \delta)$ such that:

$$\int_{S \backslash B(p,\delta)} |K(r,p)|^2 \, dS(r) \leq K(r_o, p) < +\infty \, .$$

We integrate this relation with respect to the variable p over the enclosure surface S and formally proceed to the limit by letting $\delta \to 0$. It follows immediately, since the measure $mes(B(p, \delta))$ approaches zero as δ decreases, that:

$$\int_S \int_S |K(r,p)|^2 \, dS(r) dS(p) \leq 1 \, .$$

We have the equality if the enclosure surface S is sufficiently smooth.

As immediate outcome of the square integrable character of $K(r, p)$, we stipulate that the integral transform with kernel $K(r, p)$ is a Hilbert–Schmidt integral operator[2] over the space of square integrable functions. Hilbert–Schmidt integral operators are completely continuous or compact operators.

---

[2]We merely recall that a kernel $N(r, p)$ defined over a fundamental domain $S \times S$ and for which

$$\int_S \int_S |N(r,p)|^2 \, \mu(dS(r))\mu(dS(p)) < +\infty$$

is a Hilbert–Schmidt kernel and the corresponding integral transform defined on the set of square integrable functions is known as the Hilbert–Schmidt integral operator.

The space $L^2(S)$ of all equivalence classes of real-valued Lebesgue measurable functions $\varphi(p)$, defined almost everywhere over the interior surface S, whose square $|\varphi(p)|^2$ are Lebesgue-integrable over S, introduced hereabove is a Hilbert space. This space is a generalization to the continuous variables of Euclidean spaces in finite dimension. In mathematical terms, $L^2(S)$ is written as:

$$L^2(S) = \left\{ \varphi : S \to \mathbb{R} \; measurable; \quad \int_S |\varphi(p)|^2 \, \mu(dS(p)) < +\infty \right\}.$$

The inner product on $L^2(S)$ is defined by the relation

$$\forall \varphi, \psi \in L^2(S), \quad (\varphi, \psi)_{L^2(S)} = \int_S \varphi(p)\psi(p)\mu(dS(p)) \,,$$

and the associated norm by

$$\forall \varphi \in L^2(S), \quad \|\varphi\|_{L^2(S)} = \left((\varphi, \varphi)_{L^2(S)} \right)^{1/2}$$

has the monotonicity property that if

$$A \subseteq B \subseteq S,$$

then

$$\|\varphi\|_{L^2(A)} \leq \|\varphi\|_{L^2(B)} \,.$$

We merely recall that a real Hilbert space is a linear vector space over the real numbers; that is, a set of elements which can be added and multiplied by real numbers in such a way that the usual laws of arithmetic hold, and on which an inner product is defined, the space being complete in the metric generated by the inner product.

With a view to physical relevance the functions to be examined so far are bounded quantities. The functional space on which a mathematical problem for the radiative heat transfer can be formulated appears to be the space $L^2(S)$ of measurable and square integrable functions over the enclosure boundary surface S. This, however, is not sufficient to formulate a satisfactory problem of mathematical physics. One needs a thorough knowledge of the features of the kernel $K(r, p)$. Henceforth, we shall enlarge our store of properties for $K(r, p)$.

The idea now is to imbed the functions considered in the set $C^o(S)$ of continuous functions defined on the fundamental domain S.

The character of further discussion on additional properties of the kernel $K(r, p)$ is motivated here by the practical reality. Nevertheless, a motivation from a mathematical point of view is due mainly to the fact that numerically we shall employ the piecewise linear Courant Element to approximate our functions. We do not, however, want to lose ourselves in an attempt at complete systematization. We merely observe:

The kernel $K(r,p)$ is defined and continuous for all $r \neq p$. Proceeding to the limit $r = p$, it approaches the value $(1/4\pi)(c(p))^2$ which in turn is bounded from above by the quantity $(1/4\pi)C_{\max}^2$. Hence $K(r,p)$ can be considered as to be a weakly singular kernel; that is, it is expressible in the form:

$$K(r,p) = \frac{A(r,p)}{\|r - p\|_{\mathbf{R}^3}^{\beta-2}} \, ,$$

with $0 < \beta \leq 2$ and $A(r,p)$ being a continuous function defined on the fundamental domain $S \times S$.

In connection with the weakly singular character of $K(r,p)$, we have the following result [34]:

Continuous Transformation

The Hilbert–Schmidt integral operator with kernel $K(r,p)$ transforms every bounded function into a continuous function.

We can infer even more about the integral transform with kernel $K(r,p)$ and we refer to the literature for a great wealth on properties of weakly singular kernels. We indicate solely [19]:

Completely Continuous Operator

The Hilbert–Schmidt integral transform with kernel $K(r,p)$ is completely continuous as an operator from $L^2(S) \cap D_c(S)$ into $C^o(S)$.

Here, $D_c(S)$ denotes the set of ideal functions (or distributions) with support compactly embedded in S, and $C^o(S)$ represents the set of continuous functions defined on S. Furthermore, we have that:

The Hilbert–Schmidt integral transform with kernel $K(r,p)$ maps $\Re(S)$, the space of Radon measures endowed with the topology $\sigma_b(\Re(S), C^o(S))$; i.e., the inductive limit of the weak topology on closed sets of $\Re(S)$, into $L^1(S)$ the space of measurable and Lebesgue integrable functions.

We may also formulate the first result above if we remark[3] that boundary integral operators, in the treatment of elliptic boundary value problems, belong to the algebra of classical pseudo-differential operators of integer orders for the domain under consideration. Because the enclosure under consideration is assumed convex, we have the following important property for $K(r,p)$ [91, 92]:

---

[3]This remark was suggested to us by Prof. Dr.-Ing. Wolfgang L. Wendland at the Mathematisches Institut A, Universität Stuttgart, Germany.

Pseudo-Differential Operator

The Hilbert–Schmidt integral transform with kernel $K(r, p)$ is a pseudo-differential operator of order -2 and, hence, maps $L^2(S)$ into $H^2(S)$ continuously.

The mapping of $L^2(S)$ into $C^o(S)$, using the integral transform with kernel $K(r, p)$, follows since Sobolev's embedding theorem $H^2(S) \hookrightarrow C^o(S)$ or Kondrasov's compact injection theorem $H^2(S) \subset^c C^o(S)$ hold true.

We merely recall that the second order Sobolev space $H^2(S)$ of measurable and square integrable functions with tangential derivatives up to and including the second order in the distribution sense also square integrable is a Hilbert space.

The enclosure boundary S possesses continuous curvatures. Let $\partial_1(\cdot)$ and $\partial_2(\cdot)$ designate independent first order differential operators on the physical space $\mathbb{R}^3$ and tangential to the surface S. Then, the space $H^2(S)$ is written to be:

$$H^2(S) = \{\varphi \in L^2(S); \quad \partial_1^\alpha \varphi, \, \partial_2^\alpha \varphi \in L^2(S), \quad |\alpha| \leq 2\} \ .$$

The inner product on $H^2(S)$ and the norm are defined respectively by:

$$(\varphi, \psi)_{H^2(S)} = (\varphi, \psi)_{L^2(S)} + \int_S \sum_{|\alpha| \leq 2} \sum_{i=1}^{2} \partial_i^\alpha \varphi(p) \partial_i^\alpha \psi(p) \mu(dS(p)) \ ,$$

$$\|\varphi\|_{H^2(S)} = ((\varphi, \varphi)_{H^2(S)})^{1/2} \ .$$

The space $H^2(S)$ is separable; that is, there exists a denumerable dense[4] subset $X(S)$ in $H^2(S)$. The Hilbert space $H^2(S)$ is defined as the completion of the space of real-valued and smooth functions with the $H^2(S)$-norm $\|\cdot\|_{H^2(S)}$. This means that if φ_1, φ_2, ..., φ_n, ..., is a sequence of positive-valued functions with continuous first derivatives and uniformly bounded $H^2(S)$-norm and if, furthermore, the functions φ_i converge in the $H^2(S)$-norm (i.e. $\|\varphi_n - \varphi_m\|_{H^2(S)} \to 0$ for $n, m \to 0$) then, we attribute to the sequence $\{\varphi_i\}$ an ideal element $\varphi \equiv \varphi_\infty$ as a limit. In much the same way one introduces real numbers by completion of the system of rational numbers. The norm of the limit function is defined by:

$$\|\varphi\|_{H^2(S)} = \lim_{n \to +\infty} \|\varphi_n\|_{H^2(S)} \ .$$

To end this brief analysis of the boundary value equation, we would conjecture without loss of generality that the enclosure surface emissivity

---

[4]We say that a subset $X(S) \subset H^2(S)$ is dense if its closure is equal to $H^2(S)$; that is, it has the property that every function which can be approximated arbitrarily well in the mean (uniform convergence for the $H^2(S)$-norm) by a finite number of functions of $X(S)$ can also be approximated in the mean by functions taken from any infinite subset of $H^2(S)$.

$\varepsilon(r)$ and the medium absorption coefficient $a(p)$ are continuous functions. In working a practical furnace situation, however, physically meaningful discontinuity of emissivity may occur across surfaces made of different materials. Nevertheless, when such situations occur, the speckled wall in the furnace (e.g. cooling pipes and refractory bricks) can be replaced with an equivalent surface assumed to be homogeneous (e.g. the equivalent uniform plane concept used by Hottel [78]) with a continuous emissivity. The assumption of continuous surface emissivity is therefore plausible.

Our considerations show that if a solution $q(p)$ to the boundary value equation exists it must be a continuous function defined over the interior surface S. For convex enclosure geometry, the solution $q(p)$, if it exists, belongs to $H^2(S)$. We may restrict ourselves to piecewise continuous functions; that is, we may consider functions for which the fundamental domain may be subdivided into a finite number of domains that, in the interior of each domain, the function is continuous and approaches a finite limit as a point on the boundary of one of these domains is approached from its interior.

This restriction may be of utility since a piecewise continuous function is uniquely determined by its expansion coefficients with respect to a given complete[5] system of functions. Therefore, it can be approximated as closely as desired, in the sense of the quadratic Hilbert norm, by a continuous function.

After these preparations we are now in a position to formulate a properly posed mathematical problem for heat transfer by radiation in enclosures. As a result of the foregoing analysis, a functional space candidate over which our radiation problem can be formulated is the space of continuous functions.

We shall keep in mind that the mathematical problem which is to correspond to physical reality should satisfy the following basic requirements: the solution must exist; the solution should be uniquely determined; the solution should depend continuously on the data (requirement of stability). The first requirement expresses the logical condition that not too much, i.e. no mutually contradictory properties, is demanded of the solution. The second requirement stipulates completeness of the problem, leeway or ambiguity should be excluded unless inherent in the physical situation describing the radiative heat transfer. The third requirement, particularly incisive, is necessary since the mathematical formulation is to describe a physical phenomena. The next section outlines the mathematical formulation of the

---

[5]We call a system of functions defined over S complete if every piecewise continuous function, for which the integral of its square with respect to the Lebesgue measure exists, can be approximated in the mean (uniform convergence for the quadratic Hilbert norm) arbitrarily closely by a linear combination of functions of the system. This implies that the system cannot be enlarged by using convergence under the same norm.

radiative heat transfer problem in which the conditions of continuity hence square integrability prevail.

4.3 Canonical Formulation

In investigating the boundary value equation of radiative heat transfer under the hypothesis of known temperature distributions within the enclosure volume and over its boundary, we pose the following free boundary value problem which we regard as the canonical formulation:

Determine the net radiative heat flux $q(p)$, a function defined and continuous over the enclosure boundary S, such that:

$$\frac{q(p)}{\varepsilon(p)} + \int_S \left(\left(e_b(r) - \rho(r)\frac{q(r)}{\varepsilon(r)} \right) \tau(r,p) + L(r,p) \right) K(r,p)dS(r) = e_b(p),$$

where $\varepsilon(r) = 1 - \rho(r)$, $e_b(r)$ and $\tau(r,p)$ are positive and continuous functions representing respectively the interior surface emissivity, the blackbody emissive power (or Planck's function) and the transmissivity of the medium; $K(r,p)$ is the positive and real-valued function defined over $S \times S - \{r = p\}$ by:

$$K(r,p) = \frac{\cos(\theta(r,p))\cos(\theta(p,r))}{\pi \|r - p\|^2_{R^3}} .$$

The line integral taking into account the medium contribution is defined to be:

$$L(r,p) = \int_{(r,p)} a(u)e_b(u)\tau(u,p)dL(u) ,$$

with $a(u)$ being the medium absorption coefficient. The transmissivity of the medium is given as:

$$\tau(r,p) = \exp(-\int_{(r,p)} a(u)dL(u)) .$$

Although the equation for the radiative heat flux $q(p)$ in the problem above is usable it is not a trivial matter to derive a theoretical and a good computational solution procedure. One way of obtaining a numerical solution of this equation is by using a variant of the multi-grid iteration scheme of the second kind. In this case, the treatment owes a great deal to Picard's iteration algorithm [94].

A numerical approximation of the solution may also be obtained by using an appropriately modified Nyström method [95] or by using the Boundary Element Methods [45]. Since the kernel $K(r,p)$ is not defined for $r = p$, the traditional Nyström method cannot be applied directly but must be modified accordingly.

A derivation of the traditional Nyström method is given in Chap. 1 (Sect. 1.4.3). The method consists of: firstly, approximating the above integral equation by using a numerical integration rule and ignoring the error term of integration; secondly, replacing the sought for function $q(p)$ in the resulting equation by an approximation $q_h(p)$. This procedure gives an implicit equation for $q(p)$ which is called the Nyström extension of the primary integral equation governing the canonical formulation.

The Nyström extension can be regarded as a natural interpolation formula for $q(p)$ and it may be solved by collocation at the points p_i; that is, to require that the extension holds for the points p_i, with i varying from 1 to n. This yields a system of n equations in the n unknowns $q_h(p_i)$. The resulting equations are nonlinear and if solved yield approximation $q_h(p_i)$ to $q(p_i)$.

The boundary element method, for boundary value problems posed in mathematical physics, is based on the points collocation at the enclosure boundary [96]-[102]. Spline collocation methods are the most frequently employed for the numerical solution of a variety of integral equations. The method of collocation, as shown in the introductory chapter, is based on the idea of locally uniform correspondence between the exact and approximate solutions. To determine the n parameters $q_h(p_i)$, the method requires the error

$$e(p) = |q(p) - q_h(p)|$$

to vanish at the collocation points p_i. Of course these points must be chosen so that the resulting system of equations has a solution, say $q_h(p; p_1, \ldots, p_n)$. The success of collocation methods relies on the way in which the points p_i are chosen. The ideal collocation points are those for which the quatity $q_h(p; p_1, \ldots, p_n)$ minimizes the maximum error for all points p placed over the interior surface S. If we define the maximum error to be:

$$e(p_1, p_2, \ldots, p_n) = \max\left\{|q(p) - q_h(p; p_1, \ldots, p_n)|; \quad p \in S\right\},$$

then, as the collocation points we should take the p_i, $i = 1, \ldots, n$, for which this maximum error is minimum. Unfortunately, there are no general procedures presently available for a priori selection of points satisfying this criterion; in fact, the points p_i are usually determined by trials or practical considerations.

Considering the collocation method, we may realize that an exact matching of the solution at certain points does not ensure any boundness of the deviation between the exact and approximate solutions at other points placed over S. The approximate solution may vary considerably with positioning of the collocation points.

Convergence with spline collocation, for Fredholm integral equations of the second kind, has been demonstrated in special cases. The methods of analysis generally depend on the type of problem (strong elliptic problems) and the mesh utilized [100],[103]-[106]. The results, which are obtained in the topology of uniform convergence, are based on the concept of discrete perturbations of the identity by contractions and collectively compact operator families.

If we were concerned not with functions but with numbers our radiation problem henceforth would readily be solved. In this textbook, we are not concerned with just computing but we would like to produce an efficient and well-constructed algorithm whose correctness and errors can be assessed by stringent mathematical rules rather than mere plausibility. This is very important as it is essential that not only should the results be obtained using consistent mathematical analysis but also that a fundamental and unified theory should be constructed to make sure that individual computational runs do not become a simple series of empirical trials. On the other hand, to gain some insight as to where simplifying assumptions in radiative heat transfer calculations are at all plausible, it is necessary to have as exact a solution procedure as possible. An objection which is almost certain to be made is the following: what is the use of an exact and precise algorithm in a radiation problem whose enunciation itself forms, in the large scale industrial furnace situation, only a rough approximation to the furnace reality? To answer this objection, we would point out that an algorithm which gives very precise results in simple furnace situations is likely to give good results in the more complex industrial furnace case. With the above remarks in mind, we shall further reduce the canonical formulation into a simpler and more appropriate form for our investigation.

4.4   Irradiance Formulation

Since the emissivity at any point of the boundary surface S is not equal to zero we shall introduce in place of the net heat flux $q(p)$ the positive-valued function $H(p)$, which represents the surface irradiance, and defined by:

$$\forall p \in S, \quad H(p) = e_b(p) - \frac{q(p)}{\varepsilon(p)} .$$

Using the irradiance $H(p)$, the governing equation of the canonical formulation goes over into a linear integral equation of the third kind or a polar integral equation[6].Under the hypothesis of known temperature distribu-

---

[6]A linear integral equation of the second kind with the kernel $A(r)N(r,p)$, where $N(r,p)$ is symmetric and definite, i.e. it has only positive (or only negative) eigenvalues, and $A(r)$ is continuous function except for a finite number of jumps.

tions within the enclosure volume and over the boundary surface, we may formulate this equivalent problem as follows:

We seek to find a positive-valued function $H(p)$ defined and continuous over the boundary S, and solution of the integral equation

$$\forall p \in S, \quad H(p) - \int_S \left(1 - \varepsilon(r)\right) H(r)\tau(r,p)K(r,p)dS(r) = F(p) ,$$

where

$$F(p) = \int_S \left(\varepsilon(r)e_b(r)\tau(r,p) + L(r,p)\right) K(r,p)dS(r)$$

is a continuous function defined over the domain S; ε, e_b and τ being positive and continuous functions that represent respectively the enclosure interior surface emissivity, the blackbody emissive power (or Planck's function) and the transmissivity of the medium; K and L are the positive and real-valued functions already defined.

Remark 10 *The weakly singular kernel $K(r,p)$ is square integrable. Thus, the function $F(p)$ regarded as transform of the square integrable function*

$$r \longmapsto \varepsilon(r)e_b(r)\tau(r,p) + L(r,p)$$

by the Hilbert–Schmidt integral operator with kernel $K(r,p)$, is continuous.

The following questions arise immediately: How can the integral $F(p)$ be interpreted? what physical quantity is represented by the expression

$$F(p) = \int_S \left(\varepsilon(r)e_b(r)\tau(r,p) + L(r,p)\right) K(r,p)dS(r) \ ?$$

It is worth noting that the function $F(p)$ is called the forcing term (or the free term) of the linear integral equation. Here, $F(p)$ is continuous and strictly positive-valued. It can be interpreted physically as the total rate at which radiant energy per unit inside area arrives at location p by direct emission of radiation.

The quantity $F(p)$ is composed of two integral terms. The first is the attenuated radiant energy emitted by the interior surface S that arrives at the surface point p. The second is the radiant energy at the point p resulting from spontaneous emission in the $\overrightarrow{rp}$ directions by all the differential volume elements u along the paths (r,p), reduced by exponential attenuation between each point of emission u and the location p.

Similarly the following integral regarded as function of the variable p,

$$\int_S \left(1 - \varepsilon(r)\right) H(r)\tau(r,p)K(r,p)dS(r)$$

can be interpreted as the total rate at which the attenuated radiant energy per unit inside area arrives at the location p by direct reflection of radiation. From the above considerations, we see that the irradiance $H(p)$ is the total rate at which radiant energy per unit inside area directly arrives at the point p by combined emission and reflection of radiation.

In mathematical physics, linear integral equations with symmetric kernels have remarkable properties and it is therefore desired to reduce, if possible, the above integral equation to one with a symmetric kernel. With the exception of the case when the reflectivity $\rho(r) = 1 - \varepsilon(r)$ is constant, the integral equation governing the problem above is not symmetric. However, the function $\tau(r,p)K(r,p)$ is symmetric and definite, and the surface reflectivity $1-\varepsilon(r)$ ranges in the interval $]0,1[$. Accordingly, the integral equation governing our canonical formulation can be reduced advantageously into a similar one with a symmetric kernel either by moving into weighted spaces of admissible functions for the density measure $\rho(r)dS(r)$ with the reflectivity $\rho(r) = 1 - \varepsilon(r)$ being the weight function [107], or simply by setting the function $H_\varepsilon(p)$ defined by:

$$\forall p \in S, \quad H_\varepsilon(p) = H(p)\sqrt{1 - \varepsilon(p)}$$

as sought for function.

In the second approach, which we shall adopt, the function $H_\varepsilon(p)$ is positive-valued and must be continuous. Posing the radiation problem to be solved under the hypothesis of known temperature distributions within the enclosure volume and over its boundary surface, it comes that we need concern ourselves principally with the following problem:

Find a function $H_\varepsilon(p)$, defined and continuous over the boundary surface S, such that:

$$H_\varepsilon(p) - \int_S N(r,p)H_\varepsilon(r)dS(r) = F_\varepsilon(p) ,$$

where

$$N(r,p) = \tau(r,p)K(r,p)\sqrt{(1 - \varepsilon(r))(1 - \varepsilon(p))}$$

is obviously a weakly singular, symmetric and square integrable kernel, and

$$F_\varepsilon(p) = \sqrt{1 - \varepsilon(p)}F(p).$$

Since the kernel $N(r,p)$ is symmetric, the advantage which this transformation brings is that we can apply both Fredholm's and eigenvalues theory for completely continuous and self-adjoint operators to determine and give explicitly the form of the irradiance $H(p)$, hence of the net radiative flux $q(p)$. On the other hand, a non-symmetric kernel may or may not have eigenvalues.

In an analogous manner as with $K(r,p)$, we now merely add a number of general properties and remarks for the weakly singular and symmetric kernel $N(r,p)$.

Continuous Transformation

The Hilbert–Schmidt integral operator with kernel $N(r,p)$ transforms every bounded function into a continuous function.

Completely Continuous Operator

The Hilbert–Schmidt integral transform with kernel $N(r,p)$ is completely continuous as an operator from $L^2(S) \cap D_c(S)$ into $C^o(S)$. Furthermore, it maps $\Re(S)$, the space of Radon measures endowed with the topology $\sigma_b(\Re(S),\ C^o(S))$; i.e., the inductive limit of the weak topology on closed sets of $\Re(S)$, into $L^1(S)$ the space of measurable and Lebesgue integrable functions.

Pseudo-Differential Operator

For convex enclosures, the Hilbert–Schmidt integral transform with kernel $N(r,p)$ is a pseudo-differential operator of order -2 and, hence, maps $L^2(S)$ into $H^2(S)$ continuously.

Remark 11 *The kernel $N(r,p)$ is Lebesgue integrable and satisfies the following inequality:*

$$\forall p \in S, \quad \int_S N(r,p)dS(r) < 1 \ .$$

For otherwise there would exist a point $p_o \in S$ such that:

$$\int_S N(r,p_o)dS(r) = 1 \ .$$

Then, at the point $p_o \in S$ we must have:

$$\int_S (1-\tau(r,p_o))\sqrt{(1-\varepsilon(r))\,(1-\varepsilon(p_o))}K(r,p_o)dS(r) = 0 \ .$$

But since

$$\forall r \in S, \quad 1 > \tau(r,p_o)\sqrt{(1-\varepsilon(r))\,(1-\varepsilon(p_o))} \quad and \quad K(r,p_o) \geq 0 \ ,$$

we certainly obtain

$$\forall r \in S, \quad K(r,p_o) = 0 \ .$$

This leads to the relation

$$\int_S K(r,p_o)dS(r) = 0 \ ,$$

which contradicts Remark 6.

Remark 12 *The following estimate for the integral of the function $|N(r,p)|^2$ over the fundamental domain $S \times S - \{r = p\}$*

$$\int_S \int_S |N(r,p)|^2 \, dS(r) dS(p) < 1$$

holds true as a consequence of the reasoning in Remark 9.

Remark 13 *Let consider a square integrable function φ that does not vanish identically at the enclosure boundary. For, because of Remark 11, Cauchy–Schwarz' inequality yields:*

$$\left| \int_S N(r,p)\varphi(r)dS(r) \right|^2 < \int_S N(r,p) \, |\varphi(r)|^2 \, dS(r) \,.$$

If we integrate this inequality termwise with respect to the variable p over the boundary S and apply Fubini's theorem, since $N(r,p) \geq 0$, it follows immediately that:

$$\int_S \left| \int_S N(r,p)\varphi(r)dS(r) \right|^2 dS(p) < \int_S |\varphi(r)|^2 \, dS(r) \,.$$

Remark 14 *We define the quadratic integral form $J(\varphi)$ by the following relation:*

$$\forall \varphi \in L^2(S), \quad J(\varphi) = \int_S \int_S N(r,p)\varphi(r)\varphi(p)dS(r)dS(p) \,.$$

For, by using either Remark 12 or Remark 13, Cauchy–Schwarz' inequality applied to $J(\varphi)$ shows that the relation

$$\int_S |\varphi(p)|^2 \, dS(p) - \int_S \int_S \varphi(p)\varphi(r)N(r,p)dS(r)dS(p) > 0 \,,$$

holds for every square integrable function φ that does not vanish identically at the enclosure boundary.

The remarks above will also prove relevant in connection with existence proof of the solution of the variational problem. The inequality above is the one required with Enskog's method for solving linear integral equations having symmetric and positive definite kernels [17]. From Riesz–Fredholm theory for completely continuous (or compact) operators, the solution of our irradiance problem when it exists is ensured by the Fredholm alternative which we may formulate in the following way [17, 19, 38]:

Theorem 1 (Fredholm Alternative) *Either the equation governing the irradiance problem possesses one and only one continuous solution H_ε for the continuous function F_ε, in particular the solution $H_\varepsilon = 0$ for F_ε identically equal to zero, or the homogeneous problem associated to it possesses a finite positive number M of linearly independent solutions $H_{\varepsilon,1}$, $H_{\varepsilon,2}$, $\cdots$, $H_{\varepsilon,M}$ and the inhomogeneous problem has a continuous solution if and only if the function F_ε satisfies the M orthogonality conditions*

$$(F_\varepsilon, H_{\varepsilon,i})_{L^2(S)} = 0 \quad \forall i = 1, 2, \ldots, M \ .$$

In the later case, the solution of the irradiance formulation is determined only up to an additive linear combination $c_1 H_{\varepsilon,1} + c_2 H_{\varepsilon,2} + \cdots + c_M H_{\varepsilon,M}$. It may be determined uniquely by the following additional requirements:

$$(H_\varepsilon, H_{\varepsilon,i})_{L^2(S)} = 0 \quad \forall i = 1, 2, \ldots, M \ .$$

The positive-valued kernel $N(r,p)$ is square integrable and has the symmetry property. Therefore, the Hilbert–Schmidt integral transform with kernel $N(r,p)$ is linear, completely continuous, and self-adjoint. Hence, we are assured of the existence of eigenvalues and of an associated system of eigenfunctions.

Theoretically, the solution of the irradiance formulation is obtained by determining the eigenvalues[7] and the associated eigenfunctions of this integral operator. Furthermore, we have from the Fredholm alternative that: either the eigenvalues of the integral operator with kernel $N(r,p)$ are all different from unity, then the irradiance formulation has a unique solution; or some eigenvalues μ_i are equal to unity, then the solution fails if the requirements $(F_\varepsilon, H_{\varepsilon,i})_{L^2(S)} = 0$ are not satisfied.

Now we ask the question: Does there exist a solution H_ε to our irradiance problem? The answer is in the affirmative. Momentarily our problem is to characterize the eigenvalues and eigenfunctions, respectively, of the completely continuous and self-adjoint integral operator with kernel $N(r,p)$. The determination of these eigenvalues and associated eigenfunctions is based upon the following existence theorem:

Theorem 2 (Riesz–Fredholm) *Every integral transform with a symmetric and square integrable kernel $N(r,p)$ that does not vanish identically possesses eigenvalues and eigenfunctions. Their number is denumerably infinite if and only if the kernel is nondegenerate; that is, it cannot be expanded as a finite sum of products of functions of r and functions of p. All eigenvalues of the integral transform with a real valued, positive and symmetric kernel are positive and real.*

---

[7]Reciprocals of the quantities usually called eigenvalues in mathematical textbooks on spectral or eigenvalues and eigenfunctions theory.

Our real-valued kernel $N(r,p)$ is positive, symmetric and square integrable. For, applying both the method of separation of variables and the generalized Fourier series expansion to the kernel $N(r,p)$, we find that $N(r,p)$ is non-degenerate. By the Riesz–Fredholm theorem, the spectrum of the Hilbert–Schmidt integral transform with kernel $N(r,p)$ consists of a denumerable infinite sequence of real positive eigenvalues $0 < \mu_1 \leq \cdots \leq \mu_i \leq \cdots$ listed with due regard to multiplicity and magnitude.

The existence proof for our irradiance problem is completed once we can show that the value $\mu = 1$ can never occur among the eigenvalues of the Hilbert–Schmidt integral transform with kernel $N(r,p)$.

We should like to characterize the eigenvalues in such a way that either the least eigenvalue μ_1 is strictly greater than unity or the spectral radius of the Hilbert–Schmidt integral transform with kernel $N(r,p)$ is strictly less than unity. We shall prove:

The least eigenvalue of the Hilbert–Schmidt integral transform with kernel $N(r,p)$, which is the reciprocal of the maximum value which the quadratic integral form $J(\varphi)$ assumes under the subsidiary condition $(\varphi,\varphi)_{L^2(S)} = 1$, is strictly greater than unity.

Indeed, let $\mu > 0$ designate an eigenvalue of the Hilbert–Schmidt integral transform with kernel $N(r,p)$ and let φ be the associated eigenfunction. From the definition of an eigenvalue and associated eigenfunctions of the integral operator with kernel $N(r,p)$, we may write the relation:

$$\forall p \in S, \quad \varphi(p) = \mu \int_S \varphi(r) N(r,p) dS(r) .$$

For, if M is the maximum of the absolute value of the eigenfunction φ and since the function φ does not vanish identically, we have $M \neq 0$. Further, because of Remark 11, the following inequality holds:

$$\forall p \in S, \quad |\varphi(p)| < \mu M .$$

We remind: the Hilbert–Schmidt integral transform with kernel $N(r,p)$ is completely continuous as an operator from $L^2(S) \cap D_c(S)$ into $C^o(S)$. Then, the eigenfunction φ which is square integrable and with support compactly imbedded in S, $Supp(\varphi) \subset\subset S$, is actually a continuous function over the domain of definition S. According to the theorem of Weierstrass, the largest value M is attained on the surface S. Hence for $|\varphi| = M$ we obtain $M < \mu M$, giving the desired result: $1 < \mu$.

In the process of justifying our assertion, we have used the continuity property of the eigenfunction. Without going out of our way to use the continuity property of the functions considered we shall give an alternative justification, which is independent of this fact. By substituting the eigenfunction φ, as defined above, into the inequality of Remark 14 we obtain:

$$\frac{\mu - 1}{\mu} \|\varphi\|^2_{L^2(S)} > 0 .$$

But the eigenfunction φ does not vanish identically and the eigenvalue μ is positive since the real-valued kernel $N(r,p)$ is positive and symmetric. Thus, the eigenvalue μ and particularly the least eigenvalue of the kernel $N(r,p)$, or rather of the completely continuous and self-adjoint integral transform with kernel $N(r,p)$, is greater than unity.

Now if we invoke Fredholm's theorem for linear integral equations of the second kind [17],[19]-[24], it comes that the irradiance problem possesses a continuous, hence square integrable solution H_ε. Fredholm's alternative assures the uniqueness of such a function. As a consequence of this result and from the relation that links the auxiliary function $H_\varepsilon(p)$ and the surface irradiance $H(p)$, hence the net radiative heat flux $q(p)$ at every point p, it follows that the canonical solution, i.e. the net radiative heat flux distribution, uniquely exists. Thus we have carried out the existence and uniqueness proof for the solution of radiative heat transfer problem in enclosures filled in with an absorbing-emitting medium of any temperature distribution and any material properties.

The main assumption under which the solution has been established, may be summarized as follows: *The enclosure boundary S possesses continuous curvatures.* Violation of this condition will vitiate the derivation of net radiant energy balance at the surface hence of our irradiance formulation. For non-smooth boundary, however, we merely extend our results by assuming that:

Extension for a Non-smooth Boundary

The interior surface S can be subdivided by a finite number of surfaces, which possess continuous curvatures, into a finite number of sub-regions in such a way that the inner normal varies continuously in the interior of each sub-region. As the boundary of such a sub-region is approached from the interior, the inner normal tends to definite continuous values. The limiting values obtained as we approach the point on a curve separating two sub-regions may differ accordingly as we approach the point from one sub-region or the other. The inner normal over the curve that separates two sub-regions we shall then define as the arithmetic mean of the limiting values.

For, because we use Lebesgue integration theory, every statement we have made in the foregoing sections can also be extended without further complications or the introduction of new ideas to functions with jump discontinuities (e.g. the enclosure boundary emissivity).

Remark 15 *If we apply the method of successive approximations [17] to the integral equation governing the irradiance formulation we shall obtain the solution in the form of a Neumann series. This series evidently converges (for the quadratic Hilbert norm) inside a certain circle $|\mu| < R$ if the power series representing the resolvent converges in this circle. It is* *known from general theorems in the theory of functions of a complex variable that this series converges inside the circle $|\mu| < |\mu_1|$, where μ_1 is the eigenvalue, of the compact and self-adjoint integral transform with kernel $N(r, p)$, with the smallest modulus. From the foregoing developments we deduce the following result for the irradiance formulation: The method of successive approximations applied to the integral equation of our irradiance formulation converges (for the quadratic Hilbert norm) in the unit circle.*

From the developments given above we may also derive the following result for solving integral equations with symmetric kernels:

Remark 16 *Let Ω be a finite domain of $\mathbb{R}^3$ where the element of Lebesgue measure is denoted by $d\Omega(p)$. We consider a weakly singular, symmetric, positive-valued and square integrable kernel $N(r, p)$ on $\Omega \times \Omega$, the integral of which with respect to one variable is strictly less than 1; that is, a kernel for which the inequality*

$$\forall p \in \Omega, \quad \int_\Omega N(r, p) d\Omega(p) < 1$$

is valid. Then, the integral equation

$$\forall p \in \Omega, \quad \varphi(p) - \int_\Omega \varphi(r) N(r, p) d\Omega(p) = f(p)$$

does possess one and only one continuous solution $\varphi(p)$ for each arbitrary continuous function $f(p)$.

Throughout the remaining of this book, the solution H_ε of the irradiance formulation will be referred to as the Fredholm solution. In discussing our irradiance formulation, an essential point considered is the condition of Lebesgue square integrability of the sought for function $H_\varepsilon(p)$. We have made this condition more stringent by requiring, with a view to practical applications and computer simulations, continuity of the functions which are examined. The field in which the solution is to be sought is, of course, thereby restricted. This restriction, however, does not affect the solution; that is, the function which is most favorable when the wider set $L^2(S)$ of admissible functions is available is always present in the more restricted set $C^o(S)$ of continuous functions defined on the enclosure boundary S.

4.5   Analytical Form of the Solution

This section is concerned with the analytical form of the solution of heat transfer by radiation in enclosures filled in with absorbing-emitting medium. The results presented here are described in mathematical textbooks dealing with integral equations [17],[19]-[24]. Their application in radiative heat transfer is new. In the present treatise, the working equation is the governing equation of the irradiance formulation. As mentioned already in the previous section, the explicit solution is obtained by determining the eigenvalues μ_i and the associated eigenfunctions ϕ_i of the completely continuous (or compact) and self-adjoint integral transform with kernel $N(r,p)$ already defined. As a definition we merely recall that:

Definition 1 *A value μ for which the following homogeneous linear integral equation*

$$\forall p \in S, \quad \phi(p) = \mu \int_S \phi(r)N(r,p)dS(r)$$

possesses nonvanishing solutions is called an eigenvalue of the symmetric kernel $N(r,p)$ or rather of the linear, completely continuous (or compact) and self-adjoint integral transform with kernel $N(r,p)$. The corresponding solutions $\phi_1, \phi_2, \cdots, \phi_M$, assumed normalized and mutually orthogonal are called the eigenfunctions of the kernel for the eigenvalue μ. Their number is finite for each eigenvalue.

Remark 17 *An upper bound for the number M of eigenfunctions of the kernel $N(r,p)$ for the eigenvalue μ is established firstly by applying Bessel's inequality to the kernel $N(r,p)$ and the orthogonal functions $\{\phi_i\}_{i=1}^M$. The result reads:*

$$\sum_{i=1}^M |\phi_i(p)|^2 \le \mu^2 \int_S |N(r,p)|^2 \, dS(r) .$$

Secondly, by integrating the above inequality with respect to the variable p, since the above summation runs over a finite number of positive terms, we obtain that:

$$M < \mu^2 < +\infty .$$

Thus, every eigenvalue possesses a finite multiplicity or number of linearly independent eigenfunctions.

The kernel $N(r,p)$ is a symmetric and square integrable function that does not vanish identically. In conformity with the spectral theory of completely continuous and self-adjoint operators [17]-[19], $N(r,p)$ possesses a countable infinite sequence of real eigenvalues

$$1 < \mu_1 \le \mu_2 \le \cdots \le \mu_i \le \cdots,$$

and eigenfunctions $\phi_1, \phi_2, \cdots, \phi_i, \cdots$. Moreover, the set of eigenfunctions constitutes a Hilbertian basis of the space $L^2(S)$ of square integrable functions; that is, it constitutes an orthonormal basis and a dense subset of

$L^2(S)$. From the fundamental Hilbert–Schmidt Expansion theorem [17],[22]-[24], the analytical form of the Fredholm solution reads:

$$H_\varepsilon(p) = F_\varepsilon(p) + \sum_{i=1}^{+\infty} \frac{(F_\varepsilon, \phi_i)_{L^2(S)}}{\mu_i - 1} \phi_i(p) \quad \forall p \in S .$$

Theorem 3 (Hilbert–Schmidt Expansion Theorem) *Every continuous function, which is an integral transform with symmetric and square integrable kernel of a piecewise continuous function, can be expanded in a series in the eigenfunctions of that kernel. The series converges uniformly and absolutely.*

The above expansion series for H_ε converges absolutely and uniformly. To see this we need only note that, for sufficiently large i, the relation

$$|\mu_i - 1| > \frac{|\mu_i|}{2}$$

is sure to hold since μ_i approaches infinity at the limit. Thus, except for a finite number of terms, the series

$$2 \sum_{i=1}^{+\infty} \frac{|(F_\varepsilon, \phi_i)_{L^2(S)}|}{\mu_i}$$

is a majorant of our series. The orthogonal system $\phi_1, \phi_2, \cdots, \phi_i, \cdots$, is a Hilbertian basis of the space $L^2(S)$ and the forcing term F_ε is square integrable. We obtain by the completeness relation of Pythagoras' theorem for functions space that:

$$\sum_{i=1}^{+\infty} |(F_\varepsilon, \phi_i)_{L^2(S)}|^2 = \int_S |F_\varepsilon(p)|^2 \, dS(p) < +\infty .$$

Now we apply Bessel's inequality to the orthogonal system $\{\phi_i\}_{i=1}^{+\infty}$, obtaining

$$\forall p \in S, \quad \int_S |N(r,p)|^2 \, dS(r) \geq \sum_{i=1}^{+\infty} \left(\int_S \phi_i(r) N(r,p) dS(r) \right)^2 ,$$

or

$$\forall p \in S, \quad \int_S |N(r,p)|^2 \, dS(r) \geq \sum_{i=1}^{+\infty} \frac{|\phi_i(p)|^2}{\mu_i^2} .$$

If we integrate with respect to the variable p and note that $\|\phi_i\|_{L^2(S)} = 1$ we find

$$\sum_{i=1}^{+\infty} \mu_i^{-2} < 1 ;$$

that is, the sum of the reciprocal of the square of the eigenvalues converges. Hence, because of Cauchy–Schwarz' inequality, the following relation holds

$$\sum_{i=1}^{+\infty} \left| \frac{(F_\varepsilon, \phi_i)_{L^2(S)}}{\mu_i - 1} \phi_i(p) \right| \leq 2 \, \|F_\varepsilon\|_{L^2(S)} < +\infty \, .$$

If we now substitute the series expansion for H_ε into the irradiance formulation, it is clear that the governing equation of this formulation is satisfied. The explicit form of the net radiative heat flux at any boundary point p, follows immediately as:

$$q(p) = \left(e_b(p) - \frac{1}{\sqrt{1 - \varepsilon(p)}} \left(F_\varepsilon(p) + \sum_{i=1}^{+\infty} \frac{(F_\varepsilon, \phi_i)_{L^2(S)}}{\mu_i - 1} \phi_i(p) \right) \right) \varepsilon(p) \, .$$

From this relation, one may obtain the analytical forms of the intensity of incident and outgoing radiation in any direction and the form of the radiative heat source inside furnaces filled with an absorbing-emitting medium.

So far, we have shown that the theory of heat transfer by radiation leads to an eigenvalue problem. On a theoretical basis, the eigenvalues and eigenfunctions of the Hilbert–Schmidt integral transform with kernel $N(r, p)$ can be readily determined. Practically, the computing time of the procedure for calculating these eigenvalues and the associated eigenfunctions can be excessive. However, due to the symmetry of $N(r, p)$ and the postulated properties of both the forcing term F_ε and the sought for function H_ε (F_ε and H_ε are both continuous hence square integrable over S), we can also verify the existence and uniqueness of the solution and, develop a good computational procedure for finding an approximation to H_ε by use of variational principle on a well-specified Hilbert space.

The succeeding sections discuss the application of variational principle to the irradiance formulation. The variational principle is a reliable tool both in formulating and treating numerically our integral equation. As will be shown, this principle also enables to carry out existence and uniqueness of the sought for function H_ε hence of the net radiative heat flux $q(p)$. Furthermore, it gives a mean for assessing the accuracy of the approximated solution. It needs hardly be pointed out that the eigenvalue problem considered so far can also be treated from the standpoint of the calculus of variations.

4.6   Quadratic Variational Formulation

In this section we shall study the irradiance formulation of radiative heat transfer by means of variational principle. Here, we shall state our problem in a variational form, which is an equivalent integral statement of the irradiance formulation. The functional space for our variational problem remains the space $L^2(S)$. For radiative heat transfer calculations, $L^2(S)$ will represent the set of admissible rates of incident radiant energy per unit inside area.

We take the weak formulation of the equation governing the irradiance formulation in the space $L^2(S)$, as explained in Chap. 1 (Sect. 1.4.2), obtaining the following relation:

$$(H_\varepsilon, \varphi)_{L^2(S)} - \int_S \int_S H_\varepsilon(r)\varphi(p)N(r,p)dS(r)dS(p) = (F_\varepsilon, \varphi)_{L^2(S)}.$$

This leads us to the variational problem:

Find a candidate $H_\varepsilon \in L^2(S)$ such that: for every $\varphi \in L^2(S)$, we have

$$(H_\varepsilon, \varphi)_{L^2(S)} - \int_S \int_S H_\varepsilon(r)\varphi(p)N(r,p)dS(r)dS(p) = (F_\varepsilon, \varphi)_{L^2(S)}$$

with

$$N(r,p) = \tau(r,p)K(r,p)\sqrt{(1 - \varepsilon(r))(1 - \varepsilon(p))}$$

being a positive-valued, symmetric and square integrable kernel defined over $S \times S$, and $F_\varepsilon(p)$ being the forcing term defined already.

The problem above is the quadratic variational formulation, in the space $L^2(S)$, of the irradiance formulation of radiative heat transfer. This formulation, however, has a much more physical sense and is more powerful than the irradiance or canonical one. We can assert that if the governing equation of our variational formulation is satisfied for all functions $\varphi \in L^2(S)$ then, the governing equation of the irradiance formulation must be satisfied at all points of the boundary surface S.

In the first integral term, $(H_\varepsilon, \varphi)_{L^2(S)}$, we can recognize the total rate H_ε of incident radiant energy per unit inside area. This includes the contribution of both emission and reflection of radiation acting on $\varphi(p)dS(p)$. Taking $\varphi(p)$ identically equal to unity we see that the first integral term describes the total incident radiant energy upon the interior surface S. Similarly, the second integral term

$$\int_S \int_S H_\varepsilon(r)\varphi(p)N(r,p)dS(r)dS(p)$$

describes the total attenuated incident radiant energy upon the surface S by direct reflection of radiation. The third integral term, $(F_\varepsilon, \varphi)_{L^2(S)}$, describes the total incident radiant energy upon S from both the surface direct emission and the medium spontaneous emission of radiation.

The variational principle as expressed above must be understood in a broader sense that extends beyond the traditional concepts of variational calculus. If we refer to dynamics this formulation will correspond to what is known as the d'Alembert's Principle. The different terms of the equation governing the variational formulation can be transformed into quantities that represent variations of physical invariants or functionals such as the kinetic and potential energies and correspond to the viewpoint of the traditional variational calculus. We will show in a following paragraph that the solution obtained by use of variational principle coincides with the Fredholm solution. Moreover, the numerical approximation associated with the variational solution is accurate and computationaly efficient for the boundary value equation of radiative heat transfer.

4.7   Variational Solution: Existence, Uniqueness

The problem defined hereabove is the formulation of an optimization question on a well-specified set of admissible functions in variational calculus. As already mentioned, the difficulty of the variational approach is that the solution may not exist even for problems meaningfully formulated. In function space, i.e. in a space of infinitely many dimensions, it is not in general possible to choose the domain of admissible functions as a compact set (or closed[8] set of continuous functions with bounded norm) in which a principle of points of accumulation is valid. Thus, existence of a solution of a particular variational problem cannot be taken for granted. A special existence proof is needed for the solution of each class of problems.

This section is therefore concerned with the study of the solvability (existence and uniqueness of the solution) of the variational problem formulated in the previous section. In the traditional variational approach, existence of a solution is based on the fundamental Weierstrass theorem that a continuous function of several variables, which are restricted to a finite closed domain, assumes a largest and smallest value in the interior or on the boundary of the domain. In the present approach, the question of existence and uniqueness of the variational solution is answered in the affirmative by the Lax–Milgram lemma. Lax–Milgram's lemma enumerating all the assumptions may be formulated as follows [18, 19, 38]:

Lemma 4 (Lax–Milgram) *We consider a (real) Hilbert space H with the norm $\|\cdot\|_H$, a bilinear form $B : \varphi, \psi \longmapsto B(\varphi, \psi)$ continuous over $H \times H$, i.e. there exists a constant M such that*

$$\forall \varphi, \psi \in H, \quad B(\varphi, \psi) \leq M \, \|\varphi\|_H \, \|\psi\|_H \;,$$

---

[8] we call a set of functions $X(S)$ closed if there exists no normalized function in $X(S)$ orthogonal (for the X-norm) to all the functions of the set; or if it has the property that every function $\varphi(p)$ which can be approximated arbitrarily well in the mean (uniform convergence for the X-norm) by a sequence of functions taken from $X(S)$, belongs to $X(S)$.

and coercive, i.e. there exists a constant $\alpha > 0$ such that

$$\forall \varphi \in H, \quad B(\varphi, \varphi) \geq \alpha \|\varphi\|_H^2 .$$

We also consider a linear form $L : \varphi \longmapsto L(\varphi)$ continuous over H, i.e. an element L of the topological dual H^* of H endowed with the topological norm

$$\|L\|_H = \sup \left\{ \frac{L(\varphi)}{\|\varphi\|_H}; \quad \varphi \in H, \ \varphi \neq 0 \right\} .$$

Then, the general variational problem: Find a function φ in the space H such that

$$\forall \psi \in H, \quad B(\varphi, \psi) = L(\psi)$$

possesses a unique solution. Furthermore, the solution φ satisfies the inequality

$$\|\varphi\|_H = \sup \left\{ \frac{L(\psi)}{\|\psi\|_H}; \quad \psi \in H, \ \psi \neq 0 \right\} ;$$

and the linear transformation $L \in H^* \longmapsto \varphi \in H$ is continuous.

To verify the conditions for using the above lemma, we shall consider the bilinear form B and the linear form L defined respectively over the set of functions $L^2(S) \times L^2(S)$ and $L^2(S)$ by:

$$B(\varphi, \psi) = (\varphi, \psi)_{L^2(S)} - \int_S \int_S \varphi(p)\psi(r)N(r,p)dS(r)dS(p) ,$$

$$L(\varphi) = (F_\varepsilon, \varphi)_{L^2(S)} = \int_S F_\varepsilon(p)\varphi(p)dS(p) .$$

This leaves us with the problem of finding a function H_ε in $L^2(S)$ for which the following variational equation is satisfied:

$$\forall \varphi \in L^2(S), \quad B(H_\varepsilon, \varphi) = L(\varphi) .$$

If we refer to stress analysis, the space $L^2(S)$ will represent the set of kinetically admissible displacements. The term $B(H_\varepsilon, \varphi)$ will represent the distortion work of the medium associated to a virtual displacement φ whereas $L(\varphi)$ will represent the work of external forces applied to the medium. The value H_ε is in this case the true admissible displacement for which these two works are equal for every virtual and kinetically admissible displacement φ. The physical meanings of both $B(H_\varepsilon, \varphi)$ and $L(\varphi)$ in radiative heat transfer are not yet clear.

It should be observed that the bilinear form B is symmetric. To apply Lax–Milgram's lemma we have yet to verify that B and L are continuous forms and that the bilinear form B is coercive. Put in other terms, we should establish beyond doubt that for the above variational problem a mathematical solution consistent with physical considerations is possible.

Continuity of the Forms B and L

The kernel $N(r,p)$ is square integrable. From the triangular and the Cauchy–Schwarz inequalities it is self obvious that the bilinear form B and the linear form L are continuous, since the following majorations are valid.

$$\forall \varphi, \psi \ \in \ L^2(S), \quad |B(\varphi, \psi)| \leq 2 \, \|\varphi\|_{L^2(S)} \, \|\psi\|_{L^2(S)} \; ,$$
$$\forall \varphi \ \in \ L^2(S), \quad |L(\varphi)| \leq \|F_\varepsilon\|_{L^2(S)} \, \|\varphi\|_{L^2(S)} \; .$$

Coercivity of the Bilinear Form B

To advance further in the quest for existence and uniqueness of the variational solution we must henceforth verify the coercivity, or $L^2(S)$-ellipticity, of the bilinear form B. In other words, we must verify at this level that there exists a constant scalar $\alpha > 0$ for which there holds the inequality

$$\forall \varphi \in L^2(S), \quad B(\varphi, \varphi) \geq \alpha \, \|\varphi\|^2_{L^2(S)} \; .$$

This can be seen directly from the fact that the surface emissivity ranges in the interval $]0, 1[$. Thus, there exists a positive number ε_o such that:

$$\forall p \in S, \quad \varepsilon(p) \geq \varepsilon_o > 0 \; ,$$

Therefore the desired coercivity inequality follows immediately with $\alpha = \varepsilon_o$.

Alternatively, we may ascertain the positive definite character, hence coercivity of the bilinear form B, at least formally, by considering the real-valued quadratic functional $G(\varphi)$ defined over $L^2(S)$ by the relation:

$$\forall \varphi \in L^2(S), \quad G(\varphi) = \frac{1}{2} B(\varphi, \varphi) - L(\varphi) \; .$$

By a functional we mean a function which depends on the entire course of one or more functions rather than on a number of discrete variables; that is, the domain of a functional is a set or space of admissible functions rather than a region of coordinate space.

The physical interpretation of the functional $G(\varphi)$ in dynamics is that it represents a total mechanical energy of a given physical system while the variational equations

$$\forall \varphi \in L^2(S), \quad B(H_\varepsilon, \varphi) = L(\varphi) \; .$$

explain the principle of virtual works involved. Actually, the virtual work concept may be translated into the language of functional analysis. In stress analysis, the quadratic form $(B(\varphi, \varphi))\,/2$ represents the distortion energy, of the medium filling up the enclosure volume, associated to a virtual displacement φ and $L(\varphi)$ represents the potential energy of external forces applied to the medium. The physical invariant $G(\varphi)$ in this case represents the potential elastic energy of the physical system.

For radiative heat transfer calculations, the physical meaning of $G(\varphi)$ is not yet clear. Still we may interpret physically the functional $G(\varphi)$ as an "*energy integral*". We use this terminology because in all cases in which the element φ can be treated as a translation of a system the quantity $G(\varphi)$ coincides, at least when the units of measurement are appropriately selected, with the potential energy of deformation of the system.

Now we shall observe: the functional $G(\varphi)$ is differentiable (with respect to the argument function). For, because the bilinear form B is symmetric, the gradient $G'(\varphi)$ for every function $\varphi \in L^2(S)$ is given by:

$$G'(\varphi) = \int_S \varphi(p)dS(p) - \int_S \int_S \varphi(r)N(r,p)dS(r)dS(p) - \int_S F_\varepsilon(p)dS(p).$$

The Euler equation $G'(H_\varepsilon) = 0$ is satisfied since H_ε is solution of the governing equation of the irradiance formulation.

The Hessien $G''(\varphi)$ of the functional $G(\varphi)$ is given by:

$$G''(\varphi) = mes(S) - \int_S \int_S N(r,p)dS(r)dS(p) \quad \forall \varphi \in L^2(S) .$$

The Hessien $G''(\varphi)$ is a constant, let say $\alpha(S, N)$, that does not depend on the function φ of $L^2(S)$. From Remark 14, we have $\alpha(S, N) > 0$. We summarize:

The quadratic functional $G(\varphi)$ is twice differentiable over $L^2(S)$, with the first derivative equal to 0 at the point H_ε and with the second derivative strictly positive. From the classical theory of quadratic functional, it is clear that the functional $G(\varphi)$ possesses a strict minimum at the point H_ε.

We consider the point H_ε where the quadratic functional $G(\varphi)$ possesses a strict minimum. H_ε is the solution of the governing equation of the irradiance formulation. For, because the bilinear form B is symmetric, we have the relation:

$$\forall \varphi \in L^2(S), \quad G(H_\varepsilon + \varphi) = G(H_\varepsilon) + \frac{1}{2}B(\varphi, \varphi) .$$

In addition, the following Taylor–Maclaurin expansion formula with integral remainder

$$\forall \varphi \in L^2(S), \quad G(H_\varepsilon + \varphi) = G(H_\varepsilon) + \frac{1}{2}(G''(\varphi)\varphi, \varphi)_{L^2(S)} ,$$

is valid since the quadratic functional $G(\varphi)$ is real-valued. It comes that:

$$\forall \varphi \in L^2(S), \quad B(\varphi, \varphi) = (\alpha(S, N)\varphi, \varphi)_{L^2(S)} = \alpha(S, N) \cdot \|\varphi\|^2_{L^2(S)} ,$$

which shows that the bilinear form B is positive definite and coercive, with the constant of coercivity $\alpha(S, N) > 0$. The quadratic functional $G(\varphi)$ is said to be of elliptic type and the bilinear form B is said to be coercive or $L^2(S)$-elliptic. Hence, the elliptic character of radiative heat transfer in enclosures, which justify the title of the book. This feature will prove relevant as simplifications arise in the case of symmetric and positive definite matrix when the numerical solution of the simultaneous equations involved is considered.

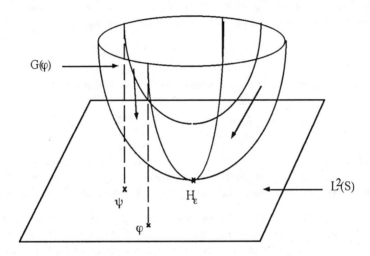

Figure 4.2. Schematic diagram of the energy integral $G(\varphi)$

Figure 4.2 above illustrates our considerations. The diagram which represents the quadratic functional $G(\varphi)$ is viewed as a paraboloidal. The horizontal sections of this paraboloidal have elliptical forms, which explains the terminology of "*elliptic type*" for the quadratic functional $G(\varphi)$.

The bilinear form B is continuous and coercive and the linear form L is continuous. From Lax–Milgram's lemma, we assert:

Our variational problem possesses a unique solution $H_\varepsilon \in L^2(S)$. The solution H_ε satisfies the inequality

$$\|H_\varepsilon\|_{L^2(S)} \le \frac{1}{\alpha(S, N)} \sup \left\{ \frac{(F_\varepsilon, \varphi)_{L^2(S)}}{\|\varphi\|_{L^2(S)}}; \quad \varphi \in L^2(S) - \{0\} \right\}$$

and is the true admissible function which realizes the minimum of the quadratic functional $G(\varphi)$; that is,

$$G(H_\varepsilon) = \min \left\{ G(\varphi); \quad \varphi \in L^2(S) \right\} .$$

4.8   Continuity of the Variational Solution

We have shown existence and uniqueness of the variational solution in the preceding section. One essential question however still remains.

To what extend does the variational solution satisfies the governing equation of the irradiance formulation?

It is trivial to see that the Fredholm solution satisfies the variational problem. Conversely, we must still verify that our variational solution is identical with the Fredholm solution. To this end, let H_ε designate the solution of our variational problem. Straightforward we have

$$\forall \varphi \in L^2(S), \quad B(H_\varepsilon, \varphi) = L(\varphi) \, .$$

Thus, in the distributions or ideal functions sense the relation

$$\forall p \in S, \quad H_\varepsilon(p) - \int_S N(r,p) H_\varepsilon(r) dS(r) = F_\varepsilon(p)$$

is valid. By using this property, in combination with the continuity of the function $F_\varepsilon(p)$ and the properties of the weakly singular kernel $N(r,p)$, we unambiguously restore the solution $H_\varepsilon(p)$ as a continuous function.

We have established that the variational solution H_ε satisfies the polar integral equation governing the irradiance formulation. By the Fredholm Alternative, H_ε is the unique solution of the irradiance problem and is a continuous function. From the developments given so far in this chapter, we conclude:

For any enclosure filled in with an absorbing-emitting medium and having opaque interior surface S with directional property effects sufficiently unimportant that it can be treated as a Lambert surface, there exists a uniquely determined radiative heat flux distribution $q(p)$ over S provided that the temperature distributions within the enclosure volume and over its interior surface are known. Furthermore, the function H_ε defined by

$$\forall p \in S, \quad H_\varepsilon(p) = \left(e_b(p) - \frac{q(p)}{\varepsilon(p)}\right)\sqrt{1 - \varepsilon(p)}$$

solves the minimum energy problem for the quadratic functional $G(\varphi)$ defined by:

$$\forall \varphi \in L^2(S), \quad G(\varphi) = \frac{1}{2} B(\varphi, \varphi) - L(\varphi) \, .$$

As immediate consequence, the intensity of incident radiation as well as the net radiative heat source and the radiation pressure distribution exist and are uniquely determined.

4.9   Question of Proper Posing Problem

We have shown that the solution H_ε uniquely exists. From a physical view point the determination of the solution H_ε requires the process of data measurement. This cannot possibly be conceived without admitting a certain amount of error. In particular, the interior surface emissivity and temperature distributions, and the medium temperature and composition are determined with a margin error. The following interrogation then arises: *to what extent is error in the measured data reflected in our variational solution?* To answer this question we stipulate:

Our variational formulation is a properly posed problem; that is, for any set of data (emissivity, temperature, medium composition) the solution H_ε is uniquely determined and for a sufficiently small (in the quadratic L^2-norm) variation of the data, there correspond an arbitrary small (in the quadratic L^2-norm) variation of the solution.

To see this, we first realize that the linear form L is continuous over $L^2(S)$. Thus, by Riesz's representation theorem that a continuous linear form $f(\varphi)$ defined on $L^2(S)$ can be represented in the form:

$$(\tau(f), \varphi)_{L^2(S)} = f(\varphi) ,$$

with

$$\tau : f \in L^2(S) \longmapsto \tau(f) \in L^2(S)$$

being Riesz's canonical isometry, there exists a unique element $l(L)$ of $L^2(S)$ such that:

$$\forall \varphi \in L^2(S), \quad L(\varphi) = (l(L), \varphi)_{L^2(S)} .$$

Similarly, if we fix a function $\varphi \in L^2(S)$ then, the linear form

$$B(\varphi, \cdot) : \psi \longmapsto B(\varphi, \psi)$$

is continuous over $L^2(S)$. Again with the aid of Riesz's representation theorem, there exists a unique element $b(\varphi)$ of $L^2(S)$ so that:

$$\forall \psi \in L^2(S), \quad B(\varphi, \psi) = (b(\varphi), \psi)_{L^2(S)} .$$

As such, we define a linear operator $b(\cdot)$ acting on $L^2(S)$. This operator is continuous since, from the continuity of the bilinear form B, the inequality

$$\|b(\varphi)\|_{L^2(S)} \leq \sup \left\{ \frac{B(\varphi, \psi)}{\|\psi\|_{L^2(S)}}; \quad \psi \in L^2(S), \ \psi \neq 0 \right\} \leq 2 \|\varphi\|_{L^2(S)}$$

holds true.

Secondly, we shall observe that our variational problem also consists of determining a function H_ε solution of the equation:

$$b(H_\varepsilon) = l(H).$$

Here, we may view the element $l(L)$ as the contribution of totality of data requires for our variational formulation. We replace the element $l(L)$ by $l(L) + \delta l(L)$ wherein $\delta l(L) \in L^2(S)$ represents a change in the data $l(L)$ and let the solution of the problem above be $H_\varepsilon + \delta H_\varepsilon$. We obtain the equation $b(\delta H_\varepsilon) = \delta l(L)$ for the change δH_ε in the solution of the variational formulation.

To show that our problem is properly posed we henceforth seek to find the inverse operator $b^{-1}(\cdot)$ and our assertion is complete once we can verify that the inverse operator exists and is continuous, hence bounded. To demonstrate this, it suffices to ascertain that the operator $b(\cdot)$ is a bijection (i.e. it defines a unique one-to-one relation) between elements of $L^2(S)$.

The coercive character of the bilinear form B implies that for every function $\varphi \in L^2(S)$ we have:

$$\alpha(S,N) \|\varphi\|^2_{L^2(S)} \le B(\varphi, \varphi) = (b(\varphi), \varphi)_{L^2(S)} \le \|b(\varphi)\|_{L^2(S)} \|\varphi\|_{L^2(S)} ;$$

that is,

$$\forall \varphi \in L^2(S), \quad \|b(\varphi)\|_{L^2(S)} \ge \alpha(S,N) \|\varphi\|_{L^2(S)} ,$$

which shows that every two elements of $L^2(S)$ with equal representation by the linear operator $b(\cdot)$ are necessarily equal.

For showing that every element of $L^2(S)$ possesses an antecedent, i.e. $b(L^2(S)) = L^2(S)$, we shall verify that the set of functions $b(L^2(S))$ is closed in $L^2(S)$ and the orthogonal $(b(L^2(S)))^\perp$ of $b(L^2(S))$ in $L^2(S)$ has the only element 0.

For this purpose let φ designate an element of the closure of $b(L^2(S))$ and let $\{b(\varphi_n)\}$ designate a sequence of elements in $b(L^2(S))$ which converges to φ. From the inequality above we write

$$\|b(\varphi_n) - b(\varphi_m)\|_{L^2(S)} \ge \alpha(S,N) \|\varphi_n - \varphi_m\|_{L^2(S)} ,$$

so that $\{b(\varphi_n)\}$ is a convergent Cauchy sequence in $L^2(S)$. Because the operator $b(\cdot)$ is continuous we obtain

$$\varphi = b(\lim_{n\to+\infty} \varphi_n) \in b(L^2(S)) .$$

Thus, $b(L^2(S))$ is closed in $L^2(S)$.

Now, let $\varphi^\perp$ designate an arbitrary element of $(b(L^2(S)))^\perp$. We have

$$\alpha(S,N) \|\varphi^\perp\|^2_{L^2(S)} \le B(\varphi^\perp, \varphi^\perp) = (b(\varphi^\perp), \varphi^\perp)_{L^2(S)} = 0 ,$$

hence $\varphi^\perp = 0$ and,

$$(b(L^2(S)))^\perp = \{0\} .$$

The linear operator $b(\cdot)$ defines a unique one-to-one relation between elements of $L^2(S)$. It comes immediately that

$$\forall \varphi \in L^2(S), \quad \left\| b^{-1}(\varphi) \right\|_{L^2(S)} \leq \frac{1}{\alpha(S,N)} \left\| \varphi \right\|_{L^2(S)} .$$

Finally, if we consider an arbitrary real number $\xi > 0$, we can find a number $\eta(\xi) = \alpha(S,N)\xi$, depending only on ξ, such that for sufficiently small variation in the data (e.g. $\|\delta l(L)\|_{L^2(S)} \leq \eta(\xi)$) we have

$$\left\| \delta H_\varepsilon \right\|_{L^2(S)} = \left\| b^{-1}(\delta l(L)) \right\|_{L^2(S)} \leq \xi ,$$

from which our assertion follows immediately.

In concluding this chapter we once more emphasize that it deals with enclosures filled in with absorbing-emitting media only. Later in Chap. 9, we shall have to consider the problem of radiative heat transfer in its full extension, including anisotropic scattering. There, we shall be able to take advantage of the foregoing developments so that we have only to make simple extensions of our arguments.

5
Numerical Approximation

In this chapter we shall be concerned with the process of approximating the exact continuous function H_ε by an expansion which behaves in a similar manner. This is an important issue for computer realization and direct applications in scientific modeling. We have shown that the solution to heat transfer by radiation in enclosures uniquely exists in a space of infinitely many dimensions. To obtain a numerical approximation of the function H_ε we shall replace our irradiance formulation in the physics of continua by an approximated one defined in a system with a finite number of degrees of freedom. It will be subsequently suitable to verify that the approximated problem has a (unique) solution and that the approximated solution converges to the true answer H_ε. This latter requirement is the main difficulty in justifying a numerical method for solving a given problem since the approximated solution do not necessarily converge to the exact solution, even if existence of the solution is not in question.

5.1 Description of the Method

An important part of the variational approximation scheme is based on the finite element method. The finite element method is a general discretization procedure of continuous problems posed by mathematically defined statements, i.e. formulations that do not consist of experimental data only. To obtain a numerical approximation of the solution H_ε we need to define a finite dimensional subspace $X_{h(n)}(S)$ of $L^2(S)$. This subspace will depend on a given size parameter $h(n)$ intended to decrease towards zero when the number n (of boundary surface elements) increases. The parameter $h(n)$ measures the coarseness or fineness of the partition of the enclosure interior surface S. At the limit when $h(n)$ decreases towards zero the dimension of $X_{h(n)}(S)$ should be infinite.

On the subspace $X_{h(n)}(S)$ we shall define an approximated problem and it would appear that solving the approximated problem in $X_{h(n)}(S)$ is equivalent to solving a linear system of equations. The solution of this system can be obtained by applying direct methods such as Gauss or Cholesky, or using relaxation and steepest descent methods.

In the purpose of determining the approximated space, the following criteria should be taken into account:

- the approximated space should ensure not only the convergence of the approximated solution, but also a good accuracy;

- The subspace should contain a basis easily assessable and such that the coefficients of the resulting linear system are simple to calculate;

- The subspace should ease the computational effort of the approximated solution.

We shall assume without loss of generality that the enclosure boundary S is polyhedral.

There do exist a large number of effective finite element types which can be used to model the subspace $X_{h(n)}(S)$ [19],[108]-[111] . In the present discussion, we use the Lagrange family of finite element defined on m-simplexes. That is, a class of elements in which the local interpolant is uniquely determined by prescribing its values at a finite number of nodal points in a region of $\mathbb{R}^m$ determined by $m + 1$ distinct points, not all of which lie in any $(m+1)$-dimensional hyperplane. These $m+1$ points define only the basic geometry of the element. A 2-simplex is a triangle, and a 3-simplex is a tetrahedron. Our analysis here follows the writings of Raviart and Thomas [18], Dautray and Lions [19], Oden and Reddy [111].

5.2   Finite Element Representation

We wish to determine an approximated space $X_{h(n)}(S)$ for the numerical solution of our variational problem. As shown in Fig. 5.1 we shall imagine a mesh $\Omega_{h(n)}(S)$ of n triangles S_k, compact[1] and connected, and with sides

---

[1]we call a set of set $\Omega(S)$,of surface elements, compact if, for every open covering C of $\Omega(S)$ there exists a finite subclass $\{C_1, C_2, \cdots, C_m\}$ of C which is an open covering of $\Omega(S)$; in other words, $\Omega(S)$ is compact if it has the property that every class of closed sets with the finite intersection property has a non empty intersection.

piecewise continuous, covering the enclosure interior surface S as follows:

$$S = \left\{ S_k \in \Omega_{h(n)}(S); \; k = 1, \cdots, n \right\} \; .$$

We require that the intersection of interior of two distinct surface elements S_i, S_j be either empty or, at most, a portion of the inter-element S_i, S_j.

The ensemble $\Omega_{h(n)}(S)$ is assumed to be a regular set of triangulation of the enclosure boundary surface S; that is, $\Omega_{h(n)}(S)$ is the set of finite surface elements (triangles) for which the ratio of radius of inscribed and circumscribed circles is uniformly bounded when the size parameter $h(n)$ approaches zero.

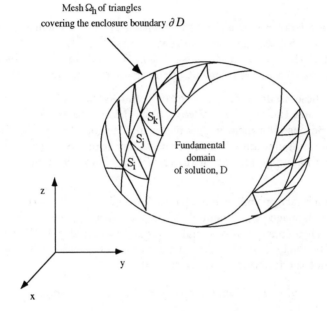

Figure 5.1. Subdivision of the enclosure boundary into finite element

We would like to choose the size parameter $h(n)$ so that the tendency of $X_{h(n)}(S)$ to $L^2(S)$ as the partition $\Omega_{h(n)}(S)$ of the boundary surface S is refined, i.e. as more surface elements of smaller dimensions are used to partition S, is represented by a tendency of $h(n)$ toward zero. To this end we take $h(n)$ as the maximum diameter of all the elements S_k:

$$h(n) = \max \left\{ \rho(S_k); \quad k = 1, \cdots, n \right\} \; .$$

The diameter $\rho(S_k)$ of the surface element S_k is defined as the maximum of Euclidean distances between two distinct points of S_k. Accordingly,

$$\rho(S_k) = \max \left\{ \|r - p\|_{\boldsymbol{R}^3} ; \quad r, p \in S_k \right\} \; .$$

Interior of each finite surface element S_k is non empty. Therefore, we consider the following ensembles:

- A finite set of m_k distinct points of S_k:

$$N(S_k) = \{p_i; \quad i = 1, \cdots, m_k\}$$

- A vectorial space $P(S_k)$ of finite dimension, formed with real-valued functions defined over S_k, and we shall consider that the vector spaces $P(S_k)$, $k = 1, 2, \cdots, n$ contain polynomials, or, at least, contain functions which are close to polynomials. The reason for this is to have direct and effective computations.

Since, for convex enclosures, the solution H_ε is sufficiently regular, in fact $H_\varepsilon \in H^2(S)$, we shall assume that $H^2(S_k) \supset P(S_k)$. We shall assume further that there is a one-to-one relation $\Re$ which links each element of $P(S_k)$ with and a unique element of the space $\mathbb{R}^{m_k}$ of $m_k - uplets$.

Under these assumptions, the set of points $N(S_k)$ is said to be $P(S_k)$-unisolvent and the triple $(S_k, P(S_k), N(S_k))$ is called the Lagrange finite element. Because the enclosure geometry may be complex, we shall identify, under a non singular affine transformation, any triple $(S_k, P(S_k), N(S_k))$ to a reference Lagrange finite element $(R, P(R), N(R))$, R being a 2-simplex (triangle).

For numerical approximation of the solution H_ε, it suffices that $P(S_k)$ be the complete space of polynomials of degree, with respect to each variable of the physical space, less or equal to a fixed integer $d \geq 1$. We shall limit ourselves to the first order polynomials space. In this case, it is not difficult to construct the functions $\varphi_i(p)$ such that:

$$\forall i = 1, \cdots, m_k; \quad \varphi_i(p_j) = \delta_{ij}, \quad 1 \leq j \leq m_k .$$

These functions, used in determining the basis of the subspace $X_{h(n)}(S)$, are fundamental for finite element analysis. In the literature they are referred to as "shape functions".

After these preliminaries we are now in a position to consider the finite elements representation of an arbitrary function $f(p)$. By finite element representation we mean an interpolant of $f(p)$ given in the form of a function, defined on the finite element being considered and which coincides with $f(p)$ at the nodal points of the finite element.

From the definition of the shape functions, we can associate to every set of m_k positive scalars $\{f_i \; ; i = 1, \cdots, m_k\}$ the real-valued function $f \in P(S_k)$ defined over S_k by the relation:

$$f(p) = f_1 \varphi_1(p) + f_2 \varphi_2(p) + \cdots + f_{m_k} \varphi_{m_k}(p) .$$

The function $f \in P(S_k)$ defined above has the important feature that: $f(p_i) = f_i$ for i varying from 1 to m_k. It is to be noted that the dimension of the vectorial space $P(S_k)$ is equal to the number m_k of points placed over the finite surface element S_k. Thus, to have the $P(S_k)$-unisolvence of the set $N(S_k)$, we need only exhibit the shape functions φ_i, $i = 1, \cdots, m_k$.

The shape functions depend on the geometry of the finite surface elements. They will be explicitly defined in a following section. Having determined these functions, the univocal relation $\Re$ is defined by:

$$\Re : \varphi \in P(S_k) \longmapsto \Re(\varphi) = (\varphi(p_i))_{i=1,\cdots,m_k} \in \mathbb{R}^{m_k}.$$

Then, to every function $f(p)$ defined over the interior surface S we associate the function $L(S_k, f)$ defined over S_k by the relation:

$$\forall p \in S_k, \quad L(S_k, f)(p) = f(p_1)\varphi_1(p) + \cdots + f(p_{m_k})\varphi_{m_k}(p),$$
$$\forall p \notin S_k, \quad L(S_k, f)(p) = 0.$$

The function $L(S_k, f)$ interpolates f at the points $p_1, p_2, \cdots, p_{m_k}$ and is called the Lagrange $P(S_k)$-interpolant of f over the surface element S_k.

The application

$$L(S_k, \cdot) : f \in P(S_k) \longmapsto L(S_k, f) \in <\varphi_1, \varphi_2, \cdots, \varphi_{m_k}> \subset \mathbb{R}^{m_k}$$

is a projection operator mapping the function f onto the m_k-dimensional space generated by the functions $\varphi_i(p)$. We shall remark that $L(S_k, f)$ is equal to f, for every $f \in P(S_k)$.

Since any triple $(S_k, P(S_k), N(S_k))$ can be identified under a non-singular affine transformation to a reference element $(R, P(R), N(R))$ we have the following estimate [18, 19],[113]:

Approximation Property

There exists a constant $C(R) > 0$, that depends only on the reference element $(R, P(R), N(R))$, such that: for every surface element $S_k \in \Omega_{h(n)}(S)$ and for every function $\varphi \in H^2(S)$ the inequality

$$\|\varphi - L(S_k, \varphi)\|_{L^2(S)} \leq C(R)(h(n))^2 |\varphi|_{H^2(S)}$$

holds true. The semi-norm $|\cdot|_{H^2(S)}$ is defined by the relation:

$$\forall \varphi \in H^2(S), \quad |\varphi|_{H^2(S)} = \sqrt{\|\varphi\|_{H^2(S)}^2 - \|\varphi\|_{L^2(S)}^2}.$$

With these results, we shall now define the subspace $X_{h(n)}(S)$.

5.3   Definition of the Approximation Space

We define the subspace $X_{h(n)}(S) \subset L^2(S)$ to be the space of continuous functions φ defined over S with restrictions $R(S_k, \varphi)$ to any finite surface element S_k pertaining in $P(S_k)$. In concise mathematical terms, $X_{h(n)}(S)$ is written as:

$$X_{h(n)}(S) = \{\varphi \in C^o(S); \quad \forall k = 1, \cdots, n, \quad R(S_k, \varphi) \in P(S_k)\} \,,$$

with $C^o(S)$ being the set of real-valued and continuous functions defined over S.

To approximate the function H_ε, it is convenient to determine a basis of the subspace $X_{h(n)}(S)$. For this purpose, we introduce the global Lagrange interpolant $L_{h(n)}(\cdot)$ defined on $C^o(S)$ and with values on $L^2(S)$ by:

$$L_{h(n)}(\varphi)(p) = L(S_1, \varphi)(p) + L(S_2, \varphi)(p) + \cdots + L(S_n, \varphi)(p) \,.$$

By a global interpolant we now mean an appropriately defined projection $L_{h(n)}(\varphi)$ of $\varphi \in L^2(S)$ onto the finite dimensional subspace $X_{h(n)}(S)$.

In the previous chapter we have searched for and found a continuous function H_ε defined over the enclosure boundary surface S for our irradiance formulation. It is however not always easy to ensure that the approximation will be continuous, which translates that the Lagrange interpolant $L_{h(n)}(\varphi)$ will satisfy the requirement of continuity between two adjacent boundary surface elements. To circumvent this difficulty we must add subsidiary conditions to our partition of the enclosure so that for every element $\varphi \in C^o(S)$, the Lagrange interpolant $L_{h(n)}(\varphi)$ belongs to $X_{h(n)}(S)$. These conditions are: the compatibility between two finite surface elements and the continuity of the Lagrange finite element.

The first condition ensures that for every two finite surface elements S_i and S_j having in common the side $S_i \cap S_j$, the restrictions $R_i(\varphi)$ and $R_j(\varphi)$, of any function $\varphi \in C^o(S)$, to the spaces $P(S_i)$ and $P(S_j)$ respectively, are equal on the inter element $S_i \cap S_j$.

The second condition means that for every boundary surface element $S_k \in \Omega_{h(n)}(S)$, the associated Lagrange finite element $(S_k, P(S_k), N(S_k))$ is such that the space $C^o(S)$ contains $P(S_k)$, and each side of S_k having m points is in univocal relation with $\mathbb{R}^m$; that is, that $(S_k, P(S_k), N(S_k))$ is of class C^o.

An immediate consequence of these subsidiary conditions is that the global Lagrange interpolant operator $L_{h(n)}(\cdot)$, defined on $C^o(S)$ and with values in $L^2(S)$, is actually with values in $C^o(S)$. Thus, we can rewrite the subspace $X_{h(n)}(S)$ in the form:

$$X_{h(n)}(S) = \{L_{h(n)}(\varphi); \quad \varphi \in C^o(S)\} \,.$$

To complete our definition and formulate the numerical problem, we need only define a basis of the subspace $X_{h(n)}(S)$.

Let consider the total set of discretization points $\{p_i; \quad i = 1, \cdots, M(n)\}$ of the Lagrange finite element. The total number $M(n)$ of points placed over the enclosure boundary surface S is equal to the dimension of the space $X_{h(n)}(S)$, and at the limit of $n \to +\infty$ we have $M(n) \to +\infty$.

Since the nodal points $\{p_j; \quad j = 1, \cdots, M(n)\}$ are distinct, the functions $\varphi_i(p)$ of $X_{h(n)}(S)$ defined over S such that:

$$\forall i = 1, \cdots, M(n); \quad \varphi_i(p_j) = \delta_{ij}, \quad 1 \le j \le M(n)$$

are linearly independent. They form a basis of the approximation space $X_{h(n)}(S)$. Therefore, every function $f \in X_{h(n)}(S)$ can be expanded at any point $p \in S$ in the form:

$$f(p) = f(p_1)\varphi_1(p) + f(p_2)\varphi_2(p) + \cdots + f(p_{M(n)})\varphi_{M(n)}(p) .$$

The coefficients $f(p_i)$ are called the *"degrees of freedom"* of the function $f(p)$. The functions φ_i are the piecewise linear Courant Elements usually referred to as shape functions. In fact, they were first suggested by Richard Courant in 1942 for approximating solutions of partial differential equations.

We shall choose the shape functions φ_i in such a way that there is a large enough number of them as to completely describe the enclosure boundary S. Without loss of generality we can assume that the function φ_i vanishes outside the surface element considered. The practical feature of this choice is to reduce its support[2]. An additional characteristic of the shape functions is that when restricted to a single finite surface element S_k, they are basis functions of the Lagrange finite element $(S_k, P(S_k), N(S_k))$.

5.4 Formulation of the Approximated Problem

The subspace $X_{h(n)}(S)$ has been well defined. From the approximation theory on Hilbert spaces, we are led to determine a function $H_{\varepsilon,h(n)}$ in $X_{h(n)}(S)$ such that the variational equation is identically satisfied in this variable and for every element φ of $X_{h(n)}(S)$, if $H_{\varepsilon,h(n)}$ is substituted in the bilinear form B; that is,

$$\forall \varphi \in X_{h(n)}(S), \quad B(H_{\varepsilon,h(n)}, \varphi) = L(\varphi) .$$

---

[2]We merely recall that the support of a function φ, for which we use the notation $Supp(\varphi)$, is the closure of the set of points p in the physical space $\mathbf{R}^3$ on which $\varphi(p) \ne 0$. Thus, the support of the shape function φ_i is the set of finite surface elements sharing common node p_i

Remark 18 *When the set of functions (φ_i) form a Hilbertian basis, i.e. when the φ_i are orthogonal and with quadratic Hilbert norm equal to unity and when the space $X_{h(n)}(S)$ is a dense subspace of $L^2(S)$, the solution $H_{\varepsilon,h(n)}$ of the approximated problem above is called the Ritz–Galerkin approximation of H_ε.*

At this stage and prior to any realization on computers it is of interest to determine how consistent the approximated formulation is. In other words, we must to determine how close the approximated problem above is to the variational problem formulated in the previous chapter.

To this end we define the lack of consistency of the numerical approximation to be:

$$C_{h(n)}(\varphi, \psi) = B(\varphi, \psi) - B(L_{h(n)}(\varphi), L_{h(n)}(\psi)) \ ,$$

and stipulate that:

The subspace $X_{h(n)}(S)$, endowed with the topology of $L^2(S)$, provides a consistent numerical approximation to the approximated problem above; that is,

$$\lim_{n \to +\infty} C_{h(n)}(\varphi, \psi) = 0 \quad \forall \varphi, \psi \in X_{h(n)}(S) \ .$$

To see this, we first note that the bilinear form B is continuous. Next, let φ and ψ designate two arbitrary elements of $X_{h(n)}(S)$, we have:

$$\left| C_{h(n)}(\varphi, \psi) \right| \le 2(C_{1,h(n)}(\varphi, \psi) + C_{2,h(n)}(\varphi, \psi) + C_{3,h(n)}(\varphi, \psi))$$

wherein

$$
\begin{aligned}
C_{1,h(n)}(\varphi, \psi) &= \left\| \varphi - L_{h(n)}(\varphi) \right\|_{L^2(S)} \left\| \psi - L_{h(n)}(\psi) \right\|_{L^2(S)} \\
C_{2,h(n)}(\varphi, \psi) &= \left\| \varphi - L_{h(n)}(\varphi) \right\|_{L^2(S)} \left\| \psi \right\|_{L^2(S)} \\
C_{2,h(n)}(\varphi, \psi) &= \left\| \psi - L_{h(n)}(\psi) \right\|_{L^2(S)} \left\| \varphi \right\|_{L^2(S)}
\end{aligned}
$$

and

$$\lim_{n \to +\infty} \left\| \varphi - L_{h(n)}(\varphi) \right\|_{L^2(S)} = \lim_{n \to +\infty} \left\| \varphi - L_{h(n)}(\varphi) \right\|_{L^2(S)} = 0 \ ;$$

which leads immediately to our assertion.

5.4.1 Solution of the Approximated Problem

We are hence seeking to find an approximated solution $H_{\varepsilon,h(n)}$ in the expansion form:

$$H_{\varepsilon,h(n)}(p) = H_{\varepsilon,h(n)}(p_1)\varphi_1(p) + \cdots + H_{\varepsilon,h(n)}(p_{M(n)})\varphi_{M(n)}(p) \ .$$

The bilinear form B defined over $X_{h(n)}(S) \times X_{h(n)}(S)$ is continuous and coercive and the linear form L defined over $X_{h(n)}(S)$ is continuous. Again, by the Lax–Milgram lemma the solution of the approximated problem above uniquely exists and solves the minimum problem:

$$G(H_{\varepsilon,h(n)}) = \min \left\{ G(\varphi); \quad \varphi \in X_{h(n)}(S) \right\} .$$

For the metric

$$\|\cdot\|_B = \sqrt{B(\cdot, \cdot)} = \sqrt{\alpha(S, N)} \, \|\cdot\|_{L^2(S)} \,,$$

the approximated function $H_{\varepsilon,h(n)}$ is the orthogonal projection of the true solution H_ε on the space $X_{h(n)}(S)$. Since the bilinear form $B(\varphi, \psi)$ is symmetric and coercive, it is immediate that solving the approximated problem is equivalent to solving the following linear system obtained from the variational equations:

$$B \cdot H_{\varepsilon,h(n)} = L \,,$$

where

$$
\begin{aligned}
H_{\varepsilon,h(n)} &\equiv \left(H_{\varepsilon,h(n)}(p_i)\right)_{1 \leq i \leq M(n)} \\
B &\equiv \left(B(\varphi_i, \varphi_j)\right)_{1 \leq i,j \leq M(n)} \\
L &\equiv \left(L(\varphi_i)\right)_{1 \leq i \leq M(n)}
\end{aligned}
$$

Indeed, the approximated variational equations must be satisfied for every function φ that belong to $X_{h(n)}(S)$, hence for every basis functions φ_i of $X_{h(n)}(S)$. The linear system is then obtained by substitution of the expansion form

$$H_{\varepsilon,h(n)}(p) = H_{\varepsilon,h(n)}(p_1)\varphi_1(p) + \cdots + H_{\varepsilon,h(n)}(p_{M(n)})\varphi_{M(n)}(p)$$

into the variational equations satisfied by the basis functions φ_i.

Alternatively, because the bilinear form $B(\varphi, \psi)$ is symmetric, we could also use a slightly different argument.

We may obtain the linear system $B \cdot H_{\varepsilon,h(n)} = L$ by differentiating the quadratic functional $G(H_{\varepsilon,h(n)})$ with respect to each of the individual degrees of freedom $H_{\varepsilon,h(n)}(p_i)$, with $i = 1, \cdots, M(n)$, and setting each result equal to zero. By differentiating $G(H_{\varepsilon,h(n)})$ in this manner and setting the differentials equal to zero, the minimum of the quadratic functional $G(H_{\varepsilon,h(n)})$ is accordingly attained for the set of values

$$\left\{ H_{\varepsilon,h(n)}(p_i) ; \quad i = 1, \cdots, M(n) \right\} .$$

Thus, the most accurate approximated solution of the assumed form

$$H_{\varepsilon,h(n)}(p) = H_{\varepsilon,h(n)}(p_1)\varphi_1(p) + \cdots + H_{\varepsilon,h(n)}(p_{M(n)})\varphi_{M(n)}(p)$$

for the variational problem, hence for the irradiance formulation is found.

We merely note that the solution of the linear system $B \cdot H_{\varepsilon,h(n)} = L$ is facilitated by its symmetry and positive definite character. Since the bilinear form $B(\varphi, \psi)$ is symmetric and coercive, the matrix B is symmetric and positive definite. This matrix is non-singular, it can then be inverted and the linear system $B \cdot H_{\varepsilon,h(n)} = L$ has a unique numerical solution. For solving the linear system $B \cdot H_{\varepsilon,h(n)} = L$, the properties of the matrix B allow use of Cholesky procedure which is based on the following theorem [35]:

Theorem 5 (Cholesky Decomposition) *If B is a symmetric and positive definite matrix then, there exists a real-valued lower triangular matrix C such that: $B = C \cdot C^t$ with C^t standing for transpose of the matrix C. Furthermore, the decomposition $B = C \cdot C^t$ may be determined uniquely by the requirement that all the diagonal elements of the matrix C be positive.*

The Cholesky procedure consists first of determining the matrix C of the decomposition $B = C \cdot C^t$ then, solving the linear systems $C \cdot Y = L$ and $C^t \cdot H_{\varepsilon,h(n)} = Y$. The elements of the matrix $C = (C_{ij})_{1 \leq i,j \leq M(n)}$ are given by the relations:

$$C_{ii} = \sqrt{B_{ii} - \sum_{k=1}^{i-1}(C_{ik})^2}\,, \quad 1 \leq i \leq M(n)\,,$$

$$C_{ji} = \frac{1}{C_{ii}}\left(B_{ij} - \sum_{k=1}^{i-1} C_{ik}C_{jk}\right), \quad 1 \leq i,j \leq M(n)\,.$$

The approximated solution $H_{\varepsilon,h(n)}$ has been uniquely determined. Now we recall the following fundamental fact already mentioned.

The meaning of a numerical solution (or numerical method), specifically for the radiative heat transfer phenomena which is governed by an integro-differential equation, is not precise unless it is supplemented by an estimate of the errors occurring; that is, unless it is accompanied by definite knowledge of the degree of accuracy attained.

It is convenient to verify that $H_{\varepsilon,h(n)}$ actually converges towards H_ε, the exact answer. The convergence of the approximated solution in the sense of the metric $\|\cdot\|_B$, which is equivalent to the quadratic $L^2(S)$-norm, is outlined in the next section by assessing the approximation error.

5.5   Convergence of the Numerical Solution

In numerical analysis, whether an approximation itself converges to the true solution is a difficult theoretical question which has to be investigated apart. In this section, we shall be concerned with the questions of how good the approximation is? and how it can be improved to approach the exact solution? The first question presumes knowledge of the exact solution which we have exhibited in the previous chapter. The second question is rational and is answered as the convergence, in the sense of the $L^2(S)$-norm, of the approximated solution.

Theoretically, one of the benefits of applying the variational principle to the irradiance formulation rely on the assessment of the approximation error. For, assuming that the enclosure boundary surface S is approximated exactly (e.g. polygonal-shape enclosures) and the integrals are evaluated exactly, we preface the proof of convergence of the numerical approximation by a simple lemma which may be formulated as follows [18]:

Lemma 6 (Cea's Lemma) *There exists a positive constant c indepen-dent of the approximated space $X_{h(n)}(S)$ such that:*

$$\left\| H_\varepsilon - H_{\varepsilon,h(n)} \right\|_{L^2(S)} \le c \cdot \inf \left\{ \left\| H_\varepsilon - \varphi \right\|_{L^2(S)} ; \quad \varphi \in X_{h(n)}(S) \right\} .$$

Since the bilinear form $B(\varphi, \psi)$ is symmetric, the above inequality is obtained for example with $c = (2/\alpha(S, N))^{1/2}$. If neither the enclosure boundary surface S is approximated exactly nor the integrals are evaluated exactly, then, the error estimate is given by Strang's lemma [19].

Céa's lemma above gives more than an optimum upper bound for the approximation error. It shows that the assessment of the magnitude of the error $\left\| H_\varepsilon - H_{\varepsilon,h(n)} \right\|_{L^2(S)}$ is an optimization issue which consists of minimizing in the space $L^2(S)$ the distance

$$d(H_\varepsilon, X_{h(n)}(S))_{L^2(S)} = \inf \left\{ \left\| H_\varepsilon - \varphi \right\|_{L^2(S)} ; \quad \varphi \in X_{h(n)}(S) \right\}$$

between the variational solution H_ε and the subspace $X_{h(n)}(S)$. The mag-nitude $\left\| H_\varepsilon - H_{\varepsilon,h(n)} \right\|_{L^2(S)}$ characterizes the degree of precision with which the approximate solution $H_{\varepsilon,h(n)}$ satisfies our variational formulation.

Using this concept of distance between a function and a space of func-tions, we shall quote that the distance between a function $\varphi \in L^2(S)$ of $L^2(S)$ and the space of functions $X_{h(n)}(S)$ is less than a positive quantity η if there exists a function of $X_{h(n)}(S)$ such that the absolute value between this function and φ is everywhere less than η.

The above distance attains its minimum at the point $H_{\varepsilon,h(n)}$, since $H_{\varepsilon,h(n)}$ is the orthogonal projection of the solution H_ε on the space $X_{h(n)}(S)$. This observation answers the question of how good the approximation is?

Of all the candidates $\varphi \in X_{h(n)}(S)$, the numerical solution $H_{\varepsilon,h(n)}$ is the closest to the actual solution H_ε in the sense of the metric $\|\cdot\|_B$.

Without more ado we assert that the function $H_{\varepsilon,h(n)}$ is in fact the best approximation which can be expected in the space of square integrable functions $L^2(S)$ for the sought for function H_ε.

We now pose the question: does the approximation $H_{\varepsilon,h(n)}$ converge to the exact solution H_ε?

Because $\Omega_{h(n)}(S)$ is a regular set of triangulations of the enclosure boundary S, this question is answered in the affirmative. We assert:

The approximation $H_{\varepsilon,h(n)}$ will yield the exact solution in the limit as the partition of the enclosure boundary surface S is refined; that is, as $h(n)$ decreases towards zero when n increases beyond all bounds.

Let us convince ourselves that the above assertion is true. The argument of this is simple. Since the expansion form is capable in the limit of reproducing exactly any continuous function conceivable then, as the solution of the approximated problem is unique it must approach in the limit of $h(n) \to 0$ the unique exact solution H_ε.

To formulate the proof conveniently we subject first the shape functions φ_i, $i = 1, 2, \cdots$, to the Gram–Schmidt process of orthogonalization and normalize the resulting system of functions. This is indeed an efficient way of obtaining a sparse matrix since the corresponding matrix $B \equiv (B(\varphi_i, \varphi_j); \quad 1 \le i, j \le M(n))$ is the identity matrix! However, due to excessive computing time which might occur, we do not recommend this process from a practical standpoint.

Next we require the resulting system of functions, also denoted by φ_i, with $i = 1, 2, \cdots$, for convenience, to form a complete set of functions of $L^2(S)$ in the sense of the metric $\|\cdot\|_B$. This requirement is obtained since the resulting functions φ_i are congruent to the eigenfunctions ϕ_i associated to the eigenvalues μ_i of the Hilbert–Schmidt integral operator with kernel $N(r, p)$, and the set of functions

$$\sqrt{\mu_i}\phi_i, \; i = 1, 2, \cdots,$$

is a complete basis of $L^2(S)$ in the sense of the metric $\|\cdot\|_B$. It is readily seen that the approximated solution is unchanged by this.

Now, let d designates the least value of the quadratic functional $G(\varphi)$,

$$d = \min \left\{ G(\varphi); \quad \varphi \in L^2(S) \right\} = -\frac{1}{2} \|H_\varepsilon\|_B^2 \ .$$

The quantity d is the exact lower bound of the quadratic functional $G(\varphi)$. Thus, for any arbitrary positive number ξ there exists a function:

$$\psi(\xi) : p \in S \longmapsto \psi(\xi)(p) \in L^2(S)$$

in $L^2(S)$ and such that:

$$d \le G(\psi(\xi)) < d + \frac{\xi}{2} \ .$$

Further, the sequence φ_i, $i = 1, 2, \cdots$, is a complete system of functions of $L^2(S)$ in the sense of the $\|\cdot\|_B$ -norm. Therefore, there follows the possibility of choosing a natural number $n = N(\xi)$ and $M(n)$ constant scalars $\alpha_i(\xi)$, $i = 1, 2, \cdots, M(n)$ such that:

$$\left\| \psi(\xi) - \left(\alpha_1(\xi)\varphi_1 + \alpha_2(\xi)\varphi_2 + \cdots + \alpha_{M(n)}(\xi)\varphi_{M(n)} \right) \right\|_B < \frac{\xi}{k} ,$$

the value of k will be chosen latter, and the following inequality holds true

$$G(\psi_{h(n)}(\xi)) - G(\psi(\xi)) < \frac{\xi}{2} ,$$

where

$$\psi_{h(n)}(\xi) = \alpha_1(\xi)\varphi_1 + \alpha_2(\xi)\varphi_2 + \cdots + \alpha_{M(n)}(\xi)\varphi_{M(n)} .$$

To see this, we write first

$$G(\psi_{h(n)}(\xi)) - G(\psi(\xi)) = \left\| \psi_{h(n)}(\xi) - H_\varepsilon \right\|_B^2 - \left\| \psi(\xi) - H_\varepsilon \right\|_B^2 ,$$

but

$$\left\| \psi_{h(n)}(\xi) - H_\varepsilon \right\|_B - \left\| \psi(\xi) - H_\varepsilon \right\|_B \le \left\| \psi_{h(n)}(\xi) - \psi(\xi) \right\|_B ,$$

hence

$$G(\psi_{h(n)}(\xi)) - G(\psi(\xi)) \le \left(\left\| \psi_{h(n)}(\xi) - H_\varepsilon \right\|_B + \left\| \psi(\xi) - H_\varepsilon \right\|_B \right) E(\psi)$$

with

$$E(\psi) = \left\| \psi_{h(n)}(\xi) - \psi(\xi) \right\|_B .$$

Next, we realize that

$$E(\psi) = \left\| \psi_{h(n)}(\xi) - \psi(\xi) \right\|_B < \frac{\xi}{k} ,$$

henceforth

$$\left\| \psi_{h(n)}(\xi) \right\|_B < \| \psi(\xi) \|_B + \frac{\xi}{k}$$

and

$$G(\psi_{h(n)}(\xi)) - G(\psi(\xi)) < \left(2 \| \psi(\xi) \|_B + 2 \| H_\varepsilon \|_B + \frac{\xi}{k} \right) \frac{\xi}{k} .$$

For obtaining the inequality

$$G(\psi_{h(n)}(\xi)) - G(\psi(\xi)) < \frac{\xi}{2} ,$$

We need only choose k such that

$$(2\,\|\psi(\xi)\|_B + 2\,\|H_\varepsilon\|_B + \frac{\xi}{k})\frac{1}{k} < \frac{1}{2}\,,$$

and it comes that

$$d \le G(\psi_{h(n)}(\xi)) \le G(\psi(\xi)) + \frac{\xi}{2} < d + \frac{\xi}{2}\,.$$

Since the approximation $H_{\varepsilon,h(n)}$ gives the value of the minimal quadratic functional which is in error, it follows that

$$d \le G(H_{\varepsilon,h(n)}) \le G(\psi_{h(n)}(\xi)) < d + \xi\,.$$

For the fixed value ξ, there exists a positive number $\eta(\xi)$ or a positive integer $n = N(\xi)$, depending only on ξ and not on the approximated solution $H_{\varepsilon,h(n)}$, such that for a sufficiently small size parameter $h(n)$ or for a sufficiently large value of n; e.g. $h(n) \le \eta(\xi)$ or $n > N(\xi)$), we have

$$d \le G(H_{\varepsilon,h(n)}) < d + \xi\,.$$

For, because ε can be chosen as we please, we let it steadily decrease and tend to zero, the corresponding value of $n = N(\xi)$ increases beyond all bounds; the size parameter $h(n)$ must then converge to 0, and it follows that:

$$\lim_{n \to +\infty} G(H_{\varepsilon,h(n)}) = \lim_{h(n) \to 0} G(H_{\varepsilon,h(n)}) = d\,.$$

Now, if we consider the sequence of approximated solutions $H_{\varepsilon,h(n)}$, with $n = 1, 2, \cdots$, where the size parameter $h(n)$ assumes the value zero upon increasing n, we have

$$\lim_{n \to +\infty} G(H_{\varepsilon,h(n)}) = d\,.$$

The sequence $H_{\varepsilon,h(n)}$, $n = 1, 2, \cdots$, is called a minimizing sequence for the quadratic functional $G(\varphi)$.

To complete the proof of convergence we write the quadratic functional $G(\varphi)$ in the form

$$G(\varphi) = \frac{1}{2}(\|\varphi - H_\varepsilon\|_B^2 - \|H_\varepsilon\|_B^2)\,.$$

Since the bilinear form $B(\varphi, \psi)$ is coercive, it is clear that

$$\inf\{G(\varphi);\quad \varphi \in L^2(S)\} = \min\{G(\varphi);\quad \varphi \in L^2(S)\} = -\frac{1}{2}\,\|H_\varepsilon\|_B^2\,,$$

and the minimizing sequence $H_{\varepsilon,h(n)}$, $n = 1, 2, \cdots$, is characterized by the relation

$$\lim_{n \to +\infty} G(H_{\varepsilon,h(n)}) = -\frac{1}{2}\,\|H_\varepsilon\|_B^2\,.$$

Consequently, we have

$$\lim_{n \to +\infty} = \frac{1}{2}(\|\varphi - H_\varepsilon\|_B^2 - \|H_\varepsilon\|_B^2) = -\frac{1}{2}\|H_\varepsilon\|_B^2 ,$$

hence

$$\lim_{n \to +\infty} \left\|H_\varepsilon - H_{\varepsilon,h(n)}\right\|_B = 0 .$$

From which our assertion follows.

It is of interest to point out that for convex enclosure geometries our approximation satisfy the optimal second order error estimate [93, 113]:

$$\left\|H_\varepsilon - H_{\varepsilon,h(n)}\right\|_{L^2(S)} \le c \cdot (h(n))^2 \, |H_\varepsilon|_{H^2(S)} .$$

A similar error estimate can be derived for non-convex enclosures.

So far, we have accepted with a view to direct numerical applications in scientific modeling that the enclosure boundary surface S is polyhedral. This hypothesis is plausible since a large number of furnaces and combustion chambers are polyhedral in shape (cylindrical or spherical furnaces may be approximated by polyhedron), and reported data for these experimental furnaces are available and can be used to test a mathematical algorithm. We shall continue with this hypothesis.

Indeed: If the enclosure surface S is not polyhedral, it can be approximated first by a polyhedral surface S_η which vertices lie on S. Then, we triangulate S_η as before and apply our developments on S_η. The radiative heat flux distribution at the surface S_η, for corresponding temperature distributions and material properties, approximates the radiative heat flux at the enclosure boundary S arbitrary closely if S_η differs arbitrarily little from S. Because the enclosure boundary S is assumed to be sufficiently smooth we have the following result [18, 19],[113]:

There exists a constant C, which depends only on the curvature of the surface S, such that:

$$\forall S_k \in \Omega_{h(n)}(S_\eta), \quad d(S_k, S) \le C \cdot (\rho(S_k))^2 .$$

Since we are using piecewise linear polynomials of degree less or equal to one, it can be shown that the results and error estimates given in this chapter still hold. We refer the reader to the foregoing references for a great wealth of this significant treatment.

5.6 Definition of Shape Functions

We are concerned in this section with an explicit definition of the basis functions (φ_i) introduced previously. For polyhedral shape enclosures, we subdivide the boundary surface S into triangular finite surface elements, and define the piecewise linear Courant Element, or shape function, $\varphi_A(p)$ over the finite surface element $T = (ABC)$ relatively to the corner point A as shown in Fig. 5.2 by:

$$\varphi_A(p) = \frac{mes(pBC)}{mes(ABC)} 1_T(p) ,$$

where the area $mes(pBC)$ of the triangle (pBC) is defined to be:

$$mes(pBC) = \sqrt{(pB \cdot pC)^2 - (\overrightarrow{pB} \cdot \overrightarrow{pC})^2} .$$

Here, we denote by pB (respectively pC) the Euclidean distance between the points p and B (respectively p and C). For numerical applications and computer simulations, we shall map the finite surface element $T = (ABC)$ onto the two dimensional simplex

$$R = \{(\xi, \eta) \in [0, 1] \times [0, 1]; \quad 0 \le \xi + \eta \le 1\} .$$

Explicitly the shape function $\varphi_A(p(\xi, \eta))$ defined over R relatively to the corner point $A(0, 0)$ is given by:

$$\varphi_A(p(\xi, \eta)) = 1 - (\xi + \eta) .$$

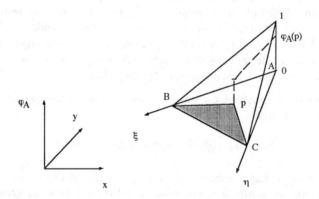

Figure 5.2. Area coordinates and geometry for definition of the linear shape function

The temperature and the material radiative properties inside the enclosure volume and over its boundary surface are prerequisite for solving the radiant energy transfer problem. The temperature distribution within the enclosure is a continuous function but may experience jump discontinuities at the boundary. Its analytical form is often unknown. Nevertheless, the temperature is known at discrete points within the enclosure. Therefore, we can approximate its distribution by using the Lagrange interpolant. The Lagrange interpolant $L_{h^m(n)}(T^m)$ of a quantity T^m is defined within the enclosure by the relation:

$$L_{h^m(n)}(T^m)(p) = L(V_1, T^m)(p) + \cdots + L(V_n, T^m)(p),$$

and for $k = 1, 2, \cdots, n$

$$L(V_k, T^m)(p) = T^m(p_1)\psi_1(p) + \cdots + T^m(p_{m_k})\psi_{m_k}(p) .$$

Here, the enclosure interior is partitioned into n finite volume elements (tetrahedral elements) V_k and $h^m(n)$ is their maximum diameter. The functions ψ_i are the basis elements of the approximated space $X_{h^m(n)}(D)$ of all admissible and real-valued functions defined within the enclosure interior. This approximated space is defined in a manner similar to $X_{h(n)}(S)$.

We also consider as shape functions ψ_i the first order polynomial functions with respect to each variable, in Cartesian coordinates, of the physical world $\mathbb{R}^3$ and defined on tetrahedrons.

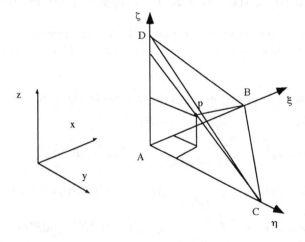

Figure 5.3. Volume coordinates and geometry for definition of the linear shape function

We merely recall that the shape function $\psi_A(p)$, first order polynomial, defined over the tetrahedron $T = (ABCD)$ relatively to the corner point A, as illustrated in Fig. 5.3, is given by:

$$\psi_A(p) = \frac{mes(pBCD)}{mes(ABCD)} 1_T(p) .$$

The volume $mes(pBCD)$ of the tetrahedron $(pBCD)$ is defined to be:

$$mes(pBCD) = \frac{1}{6}\det(\vec{1}, \overrightarrow{Op}, \overrightarrow{OB}, \overrightarrow{OC}, \overrightarrow{OD}),$$

with $\vec{1}$ being the unit vector and $O(0,0,0)$ the origin of coordinates. Over the simplex

$$R = \{(\xi, \eta, \zeta) \in [0,1] \times [0,1] \times [0,1]; \quad 0 \le \xi + \eta + \zeta \le 1\},$$

the shape function $\varphi_A(p(\xi, \eta, \xi))$ associated to the corner point $A(0,0,0)$ is given explicitly by the relation:

$$\psi_A(p(\xi, \eta, \zeta)) = 1 - (\xi + \eta + \zeta).$$

We close this chapter with considerations about a matter of general significance, the importance of which the final numerical result depends upon. The point in question is the estimation of integrals appearing in the definition of the coefficients $B(\varphi_i, \varphi_j)$ and $L(\varphi_i)$ for the linear system $B \cdot H_{\varepsilon, h(n)} = L$.

5.7   Quadrature for the Coefficients of B and L

We shall be concerned in this section with the numerical problem of computing the coefficients $B(\varphi_i, \varphi_j)$ and $L(\varphi_i)$ for the linear system

$$B \cdot H_{\varepsilon, h(n)} = L.$$

The coefficients $B(\varphi_i, \varphi_j)$ and $L(\varphi_i)$ are given by the relations:

$$
\begin{aligned}
B(\varphi_i, \varphi_j) &= B_1(\varphi_i, \varphi_j) - B_2(\varphi_i, \varphi_j), \\
B_1(\varphi_i, \varphi_j) &= \int_{Supp(\varphi_i) \cap Supp(\varphi_j)} \varphi_i(p)\varphi_j(p)dS(p), \\
B_2(\varphi_i, \varphi_j) &= \int_{Supp(\varphi_i)} (\varphi_i(p)\sqrt{1-\varepsilon(p)})B_3(p, \varphi_j)dS(p), \\
B_3(p, \varphi_j) &= \int_{Supp(\varphi_j)} \varphi_j(r)\sqrt{1-\varepsilon(r)}N(r,p)dS(r),
\end{aligned}
$$

and

$$L(\varphi_i) = \int_{Supp(\varphi_i)} \varphi_i(p)\sqrt{1-\varepsilon(p)}F(p)dS(p).$$

The term $B_1(\varphi_i, \varphi_j)$ can be calculated explicitly, and for φ_i, φ_j being first order polynomials, it is given as:

$$
\begin{aligned}
B_1(\varphi_i, \varphi_i) &= \frac{1}{6}mes(Supp(\varphi_i)), \\
B_1(\varphi_i, \varphi_j) &= \frac{1}{12}mes(Supp(\varphi_i) \cap Supp(\varphi_j)), \quad \forall i \ne j.
\end{aligned}
$$

Even with simplifications on the enclosure radiative properties, the terms $B_2(\varphi_i, \varphi_j)$ and $L(\varphi_i)$ are by no means easy to evaluate analytically. It is therefore convenient to calculate these coefficients by using quadrature formulas.

A quadrature formula for any known, definite, and integrable function φ over the finite surface element S_k is of the form[3]:

$$\int_{S_k} \varphi(r)dS(r) = \omega_1\varphi(r_1) + \omega_2\varphi(r_2) + \cdots + \omega_{m_k}\varphi(r_{m_k}) + O(h^{d+1}) \, , \quad (5.1)$$

with m_k being the number of quadrature points chosen so as to lie on the surface S_k, ω_i being the weight factors or coefficients accompanying the quadrature points r_i, and d being the quadrature order.

Any rule of approximate integration such as (5.1), no matter how the weight factors ω_i and the quadrature points r_i are chosen, will integrate exactly every function of an infinite dimensional set of admissible functions. The quality of the numerical integration rule of the form (5.1) depends on the degree of smoothness of the integrand operated upon. The smoother the integrand, the closer the approximation, and more rapid the convergence of quadrature scheme. What we do required for a quadrature rule, however, are: efficiency, reliability, and robustness. The efficiency is measured by the amount of computer time required to evaluate an integral. The reliability is measured by the magnitude of the quadrature error. The robustness is given by the ability to evaluate correctly a broad range of integrals with an occasional failure.

In partitioning the enclosure boundary into finite surface elements S_k we may regard each surface element S_k as a Cartesian product of lower dimensional elements. Thus, we can use Gauss quadrature with integration performed using Cartesian product rules. In which case the total number of points is equal to the product of the number of points used for the lower integration rules and as such exhibits a dimensional effect. Computationaly, such expressions are not the most economical and the amount of functions evaluation would be excessive. It is, however, much more desirable and aesthetically pleasing to use formulas that involved fewer points than required for a product layout. Existence theorems for such formulas and series of necessary sampling points and weights can be found in reference textbooks of numerical integration [95],[108]-[114].

With numerical integration used to substitute exact integration of the coefficients $B(\varphi_i, \varphi_j)$ and $L(\varphi_i)$, an additional error is introduced into the calculations. The first requirement is that this should be reduced as much as possible. We should be aware of the fact that it is not possible to obtain

---

[3]We use the notation according to which $O(f(x))$ denotes a function $g(x)$ for which the quotient $|g(x)/f(x)|$ remains bounded as the argument decreases.

as much accuracy with higher integration order as it is with a lower order of integration for reasonable computing times. One must naturally allow more computing time for higher orders, but we are not yet prepared to stay on the machine for months with just integral evaluations. We shall henceforth determine at once the integration order which will guarantee convergence at reasonable computing times in numerical simulations of radiant energy transfer in enclosures.

5.7.1 Required Order of Numerical Integration

We are using piecewise linear Courant Elements to approximate our functions. We shall then require that such functions be correctly integrated by a quadrature rule of the form (5.1). Thus, the minimum order of integration for convergence is one. Such a low order integration is not a realistic choice for advanced scientific modeling. It is noteworthy that to calculate the coefficients $B(\varphi_i, \varphi_j)$, a product of two shape functions needs to be evaluated. Since the shape functions are first order polynomials, both the product of two shape functions and the primitive or integral representation of a shape function are polynomials of degree, with respect to each variable of the physical space, less or equal to two.

Therefore, we shall require the quadrature method to have a degree of exactness $d \geq 2$; that is, to be exact for all polynomials of degree less or equal to d and not exact for at least one polynomial of degree $d + 1$ with $d \geq 2$. Performing the integration to accuracy order d would guarantee no loss of convergence in the variational approximation [108]. For polyhedral shape enclosures we would like not to have sampling points of integration lying on the border of any finite surface element. We shall therefore use open type integration formulas of order $d \geq 2$. Adequate values of d that make manageable the cost of numerical integration and that we shall use in our numerical simulations are $d = 2$ or $d = 3$.

5.7.2 Singular Integrals of the Kernel K(r,p)

The apparent singular behavior of the integrals defining the coefficients $B(\varphi_i, \varphi_j)$ and $L(\varphi_i)$ is observed when the kernel function $K(r, p)$ is not defined everywhere over the supports $Supp(\varphi_i)$ and $Supp(\varphi_j)$ of the basis functions φ_i and φ_j respectively. It appears when the common support $Supp(\varphi_i) \cap Supp(\varphi_j)$ is not empty. For polyhedral enclosures we merely note:

For any integrable functions $g(r, p)$ and $h(r)$ in the Lebesgue sense, the following integrals, with the already defined kernel $K(r, p)$,

$$\int_{S_k} \int_{S_k} g(r, p) K(r, p) dS(r) dS(p) \quad , \quad \int_{S_k \ni p} h(r) K(r, p) dS(r) ,$$

are equal to zero provided that the surface element S_k is plane.

Indeed, if the surface element S_k is plane, its curvatures are zero so that the kernel $K(r,p)$ is identically equal to zero for every points r and p on S_k. As a consequence of this, we also assert:

For any integrable functions $g(r,p)$ in the Lebesgue sense, the following relation

$$\int_{S_i} \int_{S_j} g(r,p)K(r,p)dS(r)dS(p) = 0$$

is equally valid provided that the surface elements S_i and S_j belong to the same plane surface.

These results represent the self-evident physical fact that any plane surface of finite measure can not irradiate itself. In other words, no radiation leaving a plane surface will strike it directly. Thus, for polyhedral enclosures the numerical problem posed by the apparent singularity of the kernel $K(r,p)$ will never arise.

A computational consequence of the above result for such enclosures is that the coefficients $B_2(\varphi_i, \varphi_j)$ need not be calculated when the supports $Supp(\varphi_i)$ and $Supp(\varphi_j)$ of non neighbor shape functions φ_i and φ_j, belong to the same hyperplane surface. The coefficients $B_2(\varphi_i, \varphi_j)$ in this case are equal to zero and the forcing term function $F(p)$ appearing in $L(\varphi_i)$ is an integral over the domain S without the hyperplane surface containing the support of φ_i.

In evaluating apparent singular integrals, with the kernel $K(r,p)$, for any enclosure geometry we may apply the above arguments provided that the enclosure boundary surface is subdivided into finite surface elements that can be regarded as plane.

5.7.3   Regular Integrals of the Kernel K(r,p)

The regular behavior of the integrals defining the coefficients $B(\varphi_i, \varphi_j)$ and $L(\varphi_i)$ is observed when the kernel function $K(r,p)$ is defined everywhere over the supports $Supp(\varphi_i)$ and $Supp(\varphi_j)$ of the basis functions φ_i and φ_j, respectively. It appears when the common support $Supp(\varphi_i) \cap Supp(\varphi_j)$ is empty. We are henceforth interested with integration over finite surface elements (triangles) S_k in a two dimensional space.

We can observe that the kernel $K(r,p)$ is obviously a low oscillatory function because of the mitigating effect of the numerator $\cos(\theta(r,p)) \cos(\theta(p,r))$.

For fixed $p \notin S_k$, the functions $\varphi_i(r)K(r,p)$ have first order polynomial-like behavior on the surface element S_k. Thus, we shall evaluate regular integrals of the kernel $K(r,p)$ over the two dimensional simplex

$$R = \{(\xi, \eta) \in [0,1] \times [0,1]; \quad 0 \le \xi + \eta \le 1\}$$

by using the following three-points formula:

$$\int_R \varphi(u)dS(u) = \frac{1}{6}(\varphi(a_{1,1}) + \varphi(a_{2,1}) + \varphi(a_{3,1})) + O(h^3) , \qquad (5.2)$$

with $a_{1,1}$, $a_{2,1}$ and $a_{3,1}$ being the points of triangular or barycentric coordinates $(2/3, 1/6, 1/6)$, $(1/6, 2/3, 1/6)$, $(1/6, 1/6, 2/3)$ relative to the corner points $(0,0)$, $(0,1)$ and $(1,0)$, respectively, as shown in Fig. 5.4 $(m = 1)$. We merely recall the definition of triangular coordinates.

Definition 2 (Triangular Coordinates) *Let consider* $(n + 1)$ *points*

$$P_j = (p_{ij})_{1 \le i \le n} \in \mathbb{R}^n , \quad 1 \le j \le n + 1$$

not situated on the same hyperplane. Then, for every point $u = (u_i)_{1 \le i \le n}$ *of* $\mathbb{R}^n$ *there do exist* $(n + 1)$ *scalars* $(\lambda_j(u))_{1 \le j \le n+1}$ *solution of the linear system*

$$\begin{aligned} p_{i,1}\lambda_1(u) + p_{i,2}\lambda_2(u) + \cdots + p_{i,n+1}\lambda_{n+1}(u) &= u_i , \quad 1 \le i \le n , \\ \lambda_1(u) + \lambda_2(u) + \cdots + \lambda_{n+1}(u) &= 1 . \end{aligned}$$

The scalars $\lambda_j(u)$ *are called the triangular or barycentric coordinates of the point* u *with respect to the points* P_j, $j = 1, \cdots, n + 1$.

For $p \in S$, the functions

$$r \longmapsto \varphi_i(r)K(r,p), \quad i = 1, 2, \cdots, M(n)$$

of the variable r have a first order polynomial-like behavior over S_k . Therefore, we shall evaluate the coefficients $B_2(\varphi_i, \varphi_j)$ which are integrals of the form

$$\int_{S_i} \varphi_i(p)\sqrt{1 - \varepsilon(p)}\psi(p)dS(p) ,$$

where

$$\psi(p) = \int_{S_k} \varphi_j(r)\sqrt{1 - \varepsilon(r)}\tau(r,p)K(r,p)dS(r) ,$$

by using the following four-points formula:

$$\int_R \varphi(u)dS(u) = \frac{1}{96}(25\sum_{i=1}^{3} \varphi(a_{i,1}) - 27\varphi(a_{0,1})) + O(h^4), \qquad (5.3)$$

with $a_{1,1}$, $a_{2,1}$, $a_{3,1}$ and $a_{0,1}$ being the points of triangular or barycentric coordinates $(1/3, 1/3, 1/3)$, $(3/5, 1/5, 1/5)$, $(1/5, 3/5, 1/5)$, $(1/5, 1/5, 3/5)$ relative to the corner points $(0,0)$, $(0,1)$ and $(1,0)$, respectively, as shown in Fig. 5.6 $(m = 1)$.

In determining the coefficients $L(\varphi_i)$, we are interested in integrals of the form

$$\int_{S_k} \varphi_i(p)\sqrt{1 - \varepsilon(p)} F(p)dS(p)$$

over finite surface elements S_k. The functions $\varphi_i(p)F(p)$ have nearly a second order polynomial-like behavior over S_k and we shall also evaluate the above integral by using the four-point rule (5.3).

The above integration formulas (Hammer's formulas), (5.2) and (5.3), are open type interpolatory rules. They are exact for all functions φ that belong to the space $P(S_k)$, with a convergence rate of order $O(h^3)$ and $O(h^4)$ respectively; that is, the quadrature error being reduced to one eighth and one sixteenth respectively for halving the size parameter h.

To further reduce the error introduced by numerical integration into the calculations, we shall subdivide every finite surface element S_k into a given number m of finite sub-surface elements over which the above quadratures will be applied. The quadrature points on the reference triangle are given in Fig.5.4 through to and including Fig.5.7 for $m = 1, 2, 3, 4, 9$. The associated three-points quadrature formula, Fig. 5.4-5.5, reads:

$$\int_T \varphi(u)dS(u) = \frac{mes(T)}{3m} \sum_{j=1}^{m} \sum_{i=1}^{3} \varphi(a_{i,j}) + O\left(\frac{h^3}{m^3}\right),$$

and the associated four-points formula, Fig. 5.6-5.7, reads:

$$\int_T \varphi(u)dS(u) = \frac{mes(T)}{48m} \sum_{j=1}^{m} (25 \sum_{i=1}^{3} \varphi(a_{i,j}) - 27\varphi(a_{0,j})) + O\left(\frac{h^4}{m^4}\right).$$

Since the relation

$$\int_S K(r,p)dS(r) = 1$$

is valid uniformly in the variable p, the number m of sub-surface elements can be chosen based on the requirement that the numerical value of this integral be allegedly correct to within the tolerance ε uniformly, i.e.

$$|Q_m(K(.,p)) - 1| \leq \varepsilon \quad \forall p \in S,$$

with $Q_m(\cdot)$ being the quadrature operator.

The three-point and the four-point quadrature rules (5.2) and (5.3) integrate the unit constant 1 exactly. Then, the compound or composite integration rules $Q_m(\cdot)$ which results from dividing the reference surface element R into m equivalent sub-surface elements and applying respectively the three-points and the four-points formulae to each of them converge to the exact value; that is,

$$\lim_{m \to +\infty} Q_m(\varphi) = \int_R \varphi(p)dS(p) .$$

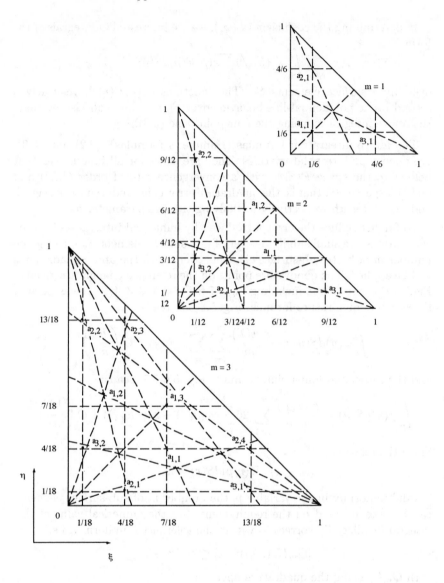

Figure 5.4. Selection of quadrature points for 3-points integration formula

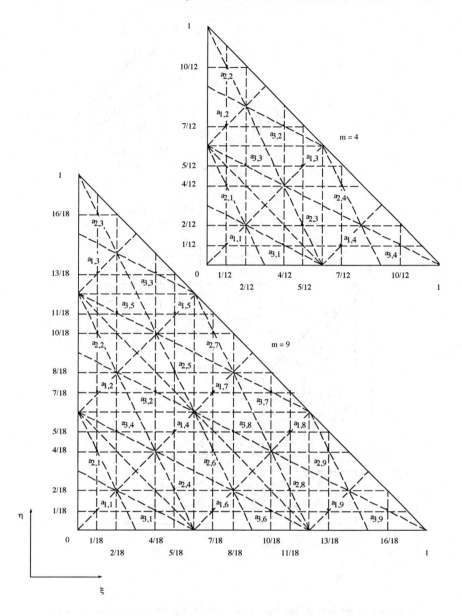

Figure 5.5. Selection of quadrature points for 3-points integration formula

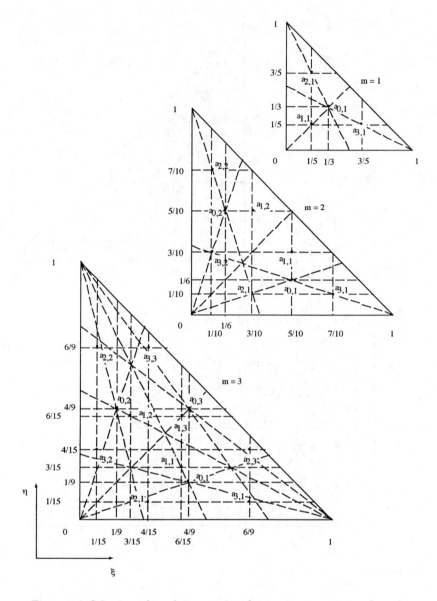

Figure 5.6. Selection of quadrature points for 4-points integration formula

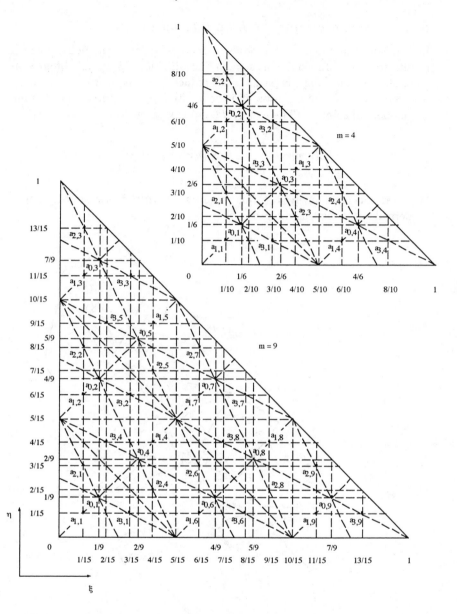

Figure 5.7. Selection of quadrature points for 4-points integration formula

5.7.4   Quadrature for the Line Integral $L(r,p)$

In non-isothermal media, the radiative energy $L(r,p)$ at the point p which results from spontaneous emission in the direction $\overrightarrow{rp}$ by all differential volume elements along the paths (r,p), reduced by exponential attenuation between each point of emission and the location p, is required to obtaining the numerical solution. The energy $L(r,p)$ is written to be:

$$L(r,p) = \int_{(r,p)} a(u)e_b(u)\tau(u,p)dL(u) .$$

We need concern ourselves in this paragraph with the way of obtaining a numerical value for this integral.

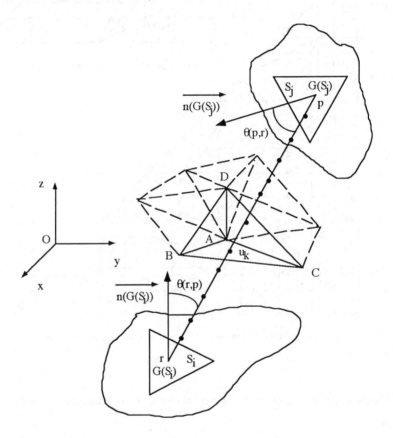

Figure 5.8. Line integral (or ray tracing) between two boundary surface elements

Merely, any primitive rule of approximate numerical integration formula, rectangular rule, midpoint rule, trapezoidal rule, etc., written in the form:

$$L(r,p) = \sum_{k=1}^{m} \omega_k a(u_k) e_b(u_k) \tau(u_k,p) + O(h^{d+1}) \,,$$

is sufficient for our purposes. Here, m is the number of quadrature points u_k running over the segment (r,p), the numbers ω_k are weight factors accompanying these points, and d is the quadrature order. To evaluate the energy $L(r,p)$ numerically, we need only determine the temperatures and/or absorption coefficients at the moving points u_k, $k = 1, \cdots, m$ inside the enclosure volume to evaluate our integral. This is a simple algorithmic question which consist of determining, for example, the tetrahedron $(ABCD)$ containing a running point u_k. Since the distributions of temperature and the coefficient of absorption inside the enclosure are known at the tetrahedron vertices, the values $T^m(u_k)$ and $a(u_k)$ are determined at once by interpolation.

A fact, which is noteworthy and useful for practical applications, is that by choosing suitably the size parameter $h(n)$ in partitioning the enclosure boundary, we need not compute all the line integrals $L(r,p)$.

Indeed: let $r \in S_i \subset S$ and $p \in S_j \subset S$ be quadrature points for surface integrations. Then, we can replace these line integrals by a weaker one, $L(G(S_i), G(S_j))$ within the tolerance $\varepsilon(mes(S_i), mes(S_j))$, which, however, runs between the center of gravities $G(S_i)$ and $G(S_j)$ of the finite surface elements S_i and S_j respectively, as illustrated in Fig. 5.8. Thus, to obtain our numerical solution, we need only compute at most $n(n+1)/2$ different line integrals (or ray tracing) in total, with n being the total number of surface elements S_k covering the enclosure boundary S.

5.8   Algorithm for Computer Realization

In this section we summarize the foregoing developments with a description of the algorithm for calculation of radiant energy exchanges in enclosures and its realization as a computer program.

Even with the actual computer technology, memory managements still remain an important factor to be considered in writing a computer program. For optimizing the use of memory space while processing, we shall store the following on the computer hard disk:

- The partition of the enclosure, the quadrature points and the radiative properties of each finite surface element use in the partition;

- The elements of both the matrix $B \equiv (B(\varphi_i, \varphi_j))_{1 \leq i,j \leq M(n)}$ and the vector $L \equiv (L(\varphi_i))_{1 \leq i \leq M(n)}$, and all intermediate matrices to be used in the calculations.

We shall also make use of specific algorithms for fast and immediate access of data stored on the computer hard disk. Then, the algorithm for radiant heat exchange calculations in enclosures, in its simple form, reads:

Algorithm: *Radiation-Exchanges*

   Begin

        {preprocessing}

- *Partition the enclosure boundary into n finite triangular surface elements*

- *Number the nodal points of the partition for ease in accessing each individual point and its neighbor points with preferably no storage*

- *Assign quadrature points for numerical integration on each finite triangular surface element*

        {processing}

- *For each surface element S_k, $k = 1, \cdots, n$ and,*
    For each surface element S_l, $l \leq k$,
      {for non-isothermal medium only}
        Compute the line integral $L(G(S_k), G(S_l))$

- *For each nodal point p_i, $i = 1, \cdots, M(n)$*
    Compute the forcing term $L(\varphi_i)$

- *For each nodal point p_i, $i = 1, \cdots, M(n)$ and,*
    For each nodal point p_j, $j = 1, \cdots, M(n)$,
      Compute the coefficient $B(\varphi_i, \varphi_j)$

- *Solve the symmetric linear system $B \cdot H_{\varepsilon,h(n)} = L$ using Cholesky method*

- *Compute from $H_{\varepsilon,h(n)}$ the vectors irradiance $H_{h(n)}$, radiosity $H_{h(n)}$ and net heat flux $q_{h(n)}$*

- *If needed, compute the net heat sources $q_{V,h(n)}$ at nodal points within the enclosure volume*

6

Simulations in Specific Cases

Quite often various conditions are imposed upon the medium filling up the enclosure volume. The medium can be held at a uniform temperature or it can have a specified absorption coefficient. The temperature could vary markedly through the medium as a result of complex turbulent combustion processes. We have shown in a preceding chapter that regardless of the medium conditions, the net radiative heat flux uniquely exists and is determined by solving an elliptic problem. The analysis has also shown that the numerical solution exists, is unique, and converges to the exact answer. In this chapter, we shall be occupied with specific cases of radiation in enclosure to ascertain both the accuracy and correctness of our algorithm.

6.1 The Transparent Medium

In a transparent medium the absorption coefficient is equal to zero, hence the radiant energy is not altered as it travels throughout the enclosure. In this case it is easy to derive the coefficients $B(\varphi_i, \varphi_j)$ and $L(\varphi_i)$, of the linear system $B \cdot H_{\varepsilon, h(n)} = L$. The forcing term $F(p)$ at any point $p \in S$ of the enclosure boundary is given as:

$$F(p) = \int_S \varepsilon(r) e_b(r) K(r, p) dS(r) .$$

If the enclosure boundary surface is sufficiently smooth (continuous curvatures) and maintained at a uniform temperature T then, the function

$$H_\varepsilon : p \longmapsto H_\varepsilon(p) = H(p)\sqrt{1 - \varepsilon(p)}$$

is the trivial solution of the irradiance formulation. As such, the following equality

$$\int_S K(r,p)dS(r) = 1, \quad \forall p \in S$$

is valid. This equality can be used to test the goodness of numerical integration. When the assumption of continuous curvatures for the surface fails, the integral

$$k(p) = \int_S K(r,p)dS(r)$$

regarded as a function of p takes values between 0 and 1. It takes the limiting value 1 if the point p is located away from the boundary surface vertex. Otherwise its value depends on the solid angle subtended by the surface S at the point p as shown already in Remark 7.

6.1.1 The Blackbody Cavity

As a first exercise, we consider the physical situation shown in Fig. 6.1. A cylindrical cavity machined in an insulated block of metal, which is uniformly maintained at a temperature $T = 500°C$, has a boundary surface covered with a gray, diffuse material of constant emissivity ε. The cavity opening is covered with a diaphragm, which has an opening of diameter $2r$. The object of this is to shield the opening from stray radiation of the surrounding and obtain a cavity that will aid in the design of black bodies for use in pyrometry and in calibrating measuring equipment. The temperature of the large surrounding room, which acts like a blackbody, is $25°C$. For numerical analysis purposes we can assume that the diaphragm opening is covered by an imaginary black surface at a temperature of $25°C$.

It is known that an optical device positioned at the opening will sense a signal proportional to the intensity leaving the bottom center of the cavity in its normal direction. Since the cavity interior surface behaves diffusely, we may define an apparent emissivity $\varepsilon_a(p) = J(p)/e_b(p)$ over it.

The apparent emissivity $\varepsilon_a(p)$ actually compares the energy leaving the location p at the cavity interior surface with the emission of a blackbody at the temperature T at the point p. The effectiveness of the blackbody cavity is henceforth measured by how close to unity the apparent emissivity is. The apparent emissivity $\varepsilon_a(p)$ may also be written in the form:

$$\varepsilon_a(p) = \varepsilon + (1 - \varepsilon)\frac{H(p)}{e_b(p)} \; .$$

where ε stands for the emissivity of the material used for manufacturing of the inner walls of the cavity.

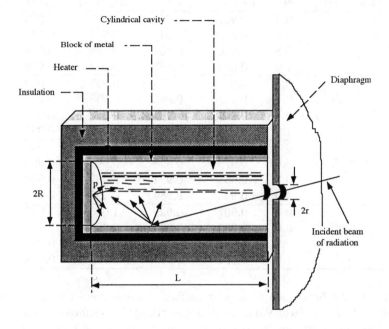

Figure 6.1. Cavity used to produce blackbody area

The objective in this first exercise is to determine how $\varepsilon_a(p)$ is related to the actual surface emissivity ε and to the dimensions (R, L, r) of the cavity. This problem is not new, it has been studied already by a number of investigators [115]-[117].

If an incident beam of radiation passes into the cavity, as illustrated above in Fig. 6.1, it strikes the cavity interior surface and part of it is absorbed with the remainder being reflected. If we make the opening very small then, the solution of our irradiance formulation approaches the value $e_b(p)$.

Thus if the opening to the cavity is sufficiently small, the apparent emissivity approaches unity. In this case the opening area behaves like the ideal black surface because essentially all radiation passing in through it is absorbed. Under such conditions, the radiation leaving the cavity interior bottom surface can be regarded as the blackbody radiation even though the surface is not a perfect emitter.

Now the question that arises is: under what geometrical conditions, determined by the three design variables (R, L, r), the opening can be regarded as sufficiently small to render apparent emissivities close to unity? We have made several computational runs to answer this question. In each case, 254

Table 6.1. Apparent emissivity at the center of the cavity bottom

Emissivity	r/R	$L/R=2$	$L/R=4$	$L/R=6$	$L/R=8$
	0.4	0.937	0.984	0.999	0.999
0.25	0.6	0.859	0.943	0.991	0.996
	0.8	0.766	0.903	0.968	0.994
	1.0	0.674	0.863	0.947	0.989
	0.4	0.975	0.995	0.998	1.000
0.50	0.6	0.943	0.982	0.993	0.999
	0.8	0.902	0.968	0.990	0.998
	1.0	0.856	0.954	0.986	0.997
	0.4	0.991	0.999	1.000	1.000
0.75	0.6	0.979	0.996	0.999	0.998
	0.8	0.963	0.994	0.999	0.998
	1.0	0.945	0.986	0.997	0.997
	0.4	0.997	0.999	1.000	1.000
0.90	0.6	0.993	0.999	0.999	0.999
	0.8	0.987	0.997	0.999	0.999
	1.0	0.981	0.995	0.998	0.998

nodal points, hence 504 triangular surface elements were placed over the cavity interior surface. We have used, per triangular surface elements, 108 quadratures points for inner integrals and 4 quadrature points for outer integrals. This resulted in approximately 7 minutes per run of a machine equipped with a Pentium II processor 400 MHz; 04 minutes for calculation of the matrix B of the linear system $B \cdot H_{\varepsilon,h(n)} = L$, 33 seconds for calculation of the vector L, and 02 minutes for solving the linear system $B \cdot H_{\varepsilon,h(n)} = L$ of size 254×254 using Cholesky method.

The numerical simulations were carried out for the cavity interior surface emissivity taking values of $\varepsilon = 0.25$, 0.50, 0.75, 0.90. In Table 6.1 we list the results for $\varepsilon_a(p)$, at the center of the cavity bottom, as a function of the diaphragm opening-to-outer radius ratio and the cavity depth-to-radius ratio. It can be observed from Table 6.1 that the apparent emissivity $\varepsilon_a(p)$ at the center of the cavity bottom increases toward unity as the cavity is made deeper and the diaphragm opening is made smaller. In Table 6.2 we list also the values obtained by other authors [115] -[116].

From Table 6.1 we may observe that the blackbody condition is achieved on the interior bottom surface when the apparent emissivity distribution over is close to unity. In this case the cavity interior bottom surface can be used as a source of blackbody energy. We observe further that the cavity length required to obtain blackbody conditions increases as the emissivity of the cavity interior surface decreases. For example, to obtain an apparent emissivity $\varepsilon_a(p) \geq 0.99$ it is necessary that $L \geq 2R$ when $\varepsilon = 0.9$, we should also have $L \geq 4R$ when $\varepsilon = 0.75$, $L \geq 6R$ when $\varepsilon = 0.50$, and $L \geq 8R$ when

Table 6.2. Apparent emissivity at the center of the cavity bottom [115]-[116]

Emissivity	r/R	$L/R = 2$	$L/R = 4$	$L/R = 8$
	0.4	0.916	0.968	0.990
0.25	0.6	0.829	0.931	0.981
	0.8	0.732	0.888	0.969
	1.0	0.640	0.844	0.965
	0.4	0.968	0.990	0.998
0.50	0.6	0.932	0.979	0.995
	0.8	0.887	0.964	0.992
	1.0	0839	0.946	0.989
	0.4	0.998	0.997	0.999
0.75	0.6	0.975	0.997	0.998
	0.8	0.958	0.988	0.997
	1.0	0.939	0.982	0.996

$\varepsilon = 0.25$. This is certainly a useful information for designing blackbody furnaces or for interpretation of pyrometer calibration measurements.

6.2 The Isothermal Gray Medium

In an isothermal medium the temperature T^m, hence the blackbody emissive power $E_b(T^m)$, does not vary in space. A medium having radiative properties that do not vary with wavelength is called a gray medium. In such medium the fractional attenuation of total radiation is independent of the nature of radiation. It is also easy to calculate the coefficients $B(\varphi_i, \varphi_j)$ and $L(\varphi_i)$, of the linear system $B \cdot H_{\varepsilon,h(n)} = L$. The forcing term $F(p)$ at any point $p \in S$ is defined as:

$$F(p) = \int_S \left(\varepsilon(r) e_b(r) \tau(r, p) + E_b(T^m)(1 - \tau(r, p)) \right) K(r, p) dS(r) ,$$

and the transmissivity $\tau(r, p)$ is defined to be:

$$\tau(r, p) = \exp(-a \, \|r - p\|_{R^3}) .$$

To illustrate the applicability of our algorithm, we pose the problem of radiative heat transfer in a well-stirred combustion chamber.

The well-stirred reactor concept is based on the assumption that mixing in the chamber is so effective that the temperature and concentration of radiating gases are uniform. Henceforth, the gas medium within the chamber can be assigned a single radiation temperature, which is the same as the furnace exit temperature.

We have designed this example to demonstrate the ability of the algorithm to predict all radiation parameters in industrial furnaces and combustion chambers which operate with sufficient momentum in the entering air and/or fuel to assure reasonably well-stirred conditions.

6.2.1   The Well-Stirred Combustion Chamber

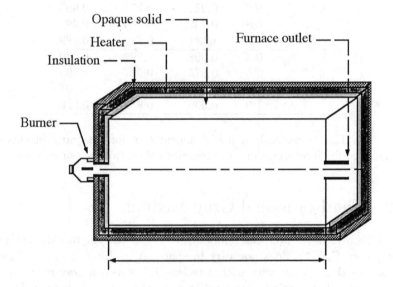

Opaque solid

Heater

Insulation

Furnace outlet

Burner

Figure 6.2. A box-shape furnace

The schematic of a rectangular enclosure of size $1m \times 1m \times 3m$ is shown in Fig. 6.2. A heat sink temperature of $1000°C$ is assigned to the furnace inner surface while a temperature of $1500°C$ is assigned to the combustion products filling the furnace volume. To calculate the heat flux distribution at the furnace interior surface we use the constant value $a = 0.1m^{-1}$ for the medium absorption coefficient. This value has been estimated by assuming that the gases products filling up the furnace volume result from combustion of methane with an excess air factor of two. The mean beam length for the enclosure under consideration is $0.77m$ (*i.e.* $3.6 \cdot Volume/Area$), which for a mixture containing 4% CO_2, 8% H_2O and 88% N_2 results in the gas total emissivity of approximately 0.085 and the absorption coefficient of $0.11m^{-1}$. The concern in this second exercise is the effect of surface emissivity on the incident heat flux $H(p)$, hence on the net heat flux to the heat sink.

The information given in this section may be used for assessing the predictions of existing radiation models. We have performed the numerical

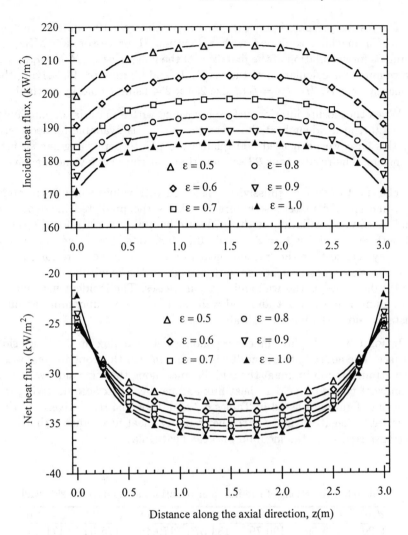

Figure 6.3. Predictions of irradiance and net radiative heat flux distributions at the furnace side wall

simulations with 254 nodal points, hence 504 triangular surface elements, placed over the furnace interior surface for each case.

For an interior surface covered with a material of emissivity $\varepsilon < 1.0$, we use 243 quadratures points for inner integrals and 4 quadrature points for outer integrals per each triangular surface elements.

The results presented hereafter took approximately 19 minutes time per run of a machine equipped with a Pentium II processor 400 MHz; 12 minutes for calculation of the matrix B of the linear system $B \cdot H_{\varepsilon,h(n)} = L$, 05 minutes for calculation of the vector L, and 02 minutes for solving the linear system $B \cdot H_{\varepsilon,h(n)} = L$ of size 254×254 using Cholesky method.

When the closure interior surface is covered with a material of emissivity $\varepsilon = 1.0$, the incident heat flux over the surface is available immediately. In this case we have used 300 quadrature points per triangular surface elements. The results took 23 seconds time per run of the same machine.

The effect of surface emissivity is assessed with values which correspond to some specific materials. Figure 6.3 shows the predicted incident flux and net flux of radiation at the enclosure boundary surface for the range of surface emissivity $\varepsilon = 0.5$, 0.6, 0.7, 0.8, 0.9, 1.0. From this figure we may realize that the incident heat flux at the furnace interior surface exhibits a symmetric behavior with respect to the axial plane $z = 1.5m$, and it decreases as the wall emissivity increases. The incident heat flux is minimum near the front and end walls and attains its maximum at mid-distance along the inner side wall.

In Fig 6.3 we display the corresponding net heat flux at the inner side surface. The net radiative heat flux is maximum near the front and end walls and attains its minimum at the mid-distance along the inner side wall. The numerical values of incident heat flux and net heat flux accompanying the foregoing figure are listed in Tables 6.3 and Table 6.4 respectively. In this later table, negative values of the net radiative heat flux indicate that heat is transferred from the medium into the heat sink.

Table 6.3. Predictions of irradiance distributions at the furnace side wall

$z(m)$	$\varepsilon = 0.5$	$\varepsilon = 0.6$	$\varepsilon = 0.7$	$\varepsilon = 0.8$	$\varepsilon = 0.9$	$\varepsilon = 1.0$
0.00	199.59	190.79	184.37	179.45	175.64	171.10
0.25	206.06	197.26	190.73	185.72	181.77	179.04
0.50	210.59	201.74	195.14	190.03	185.97	182.56
0.75	212.54	203.54	196.81	191.59	187.39	184.14
1.00	213.78	204.68	197.86	192.55	188.27	184.90
1.25	214.38	205.22	198.34	192.97	188.66	185.25
1.50	214.55	205.39	198.48	193.11	188.78	185.36
1.75	214.38	205.22	198.34	192.97	188.66	185.25
2.00	213.78	204.68	197.86	192.55	188.27	184.90
2.25	212.54	203.54	196.81	191.59	187.39	184.14
2.50	210.59	201.74	195.14	190.03	185.97	182.56
2.75	206.06	197.26	190.73	185.72	181.77	179.04
3.00	199.59	190.79	184.37	179.45	175.64	171.10

There is a close formula for calculating the net radiative heat flux at the surface of an enclosure filled in with an isothermal emitting-absorbing gas. The formula, frequently used in engineering calculation reads:

$$q = (\frac{\varepsilon_g}{\alpha_g}e_b(T_g) - e_b(T_w))/(\frac{1}{\alpha_w} + \frac{1}{\alpha_g} - 1),$$

where ε_g, α_g and $\varepsilon_w = \alpha_w$ are the gas emissivity, absorptivity and wall emissivity (absorptivity) respectively, T_g and T_w are the gas and wall (heat sink) temperatures. Applying the above formulae to the furnace under consideration and noting that $\varepsilon_g = \alpha_g$, results for in the net radiative heat flux of 35 kW/m^2 for $\varepsilon_w = 1$, which is in agreement with the results shown in Fig. 6.4.

Table 6.4. Predictions of net radiative heat flux distributions at the furnace side wall

$z(m)$	$\varepsilon = 0.5$	$\varepsilon = 0.6$	$\varepsilon = 0.7$	$\varepsilon = 0.8$	$\varepsilon = 0.9$	$\varepsilon = 1.0$
0.00	−25.32	−25.11	−24.79	−24.40	−24.02	−22.15
0.25	−28.56	−28.98	−29.25	−29.42	−29.53	−30.09
0.50	−30.82	−31.67	−32.34	−32.86	−33.31	−33.61
0.75	−31.79	−32.75	−33.50	−34.11	−34.60	−35.19
1.00	−32.42	−33.44	−34.24	−34.88	−35.39	−35.95
1.25	−32.71	−33.76	−34.57	−35.22	−35.74	−36.30
1.50	−32.80	−33.86	−34.67	−35.33	−35.84	−36.41
1.75	−32.71	−33.76	−34.57	−35.22	−35.74	−36.30
2.00	−32.42	−33.44	−34.24	−34.88	−35.39	−35.95
2.25	−31.79	−32.75	−33.50	−34.11	−34.60	−35.19
2.50	−30.82	−31.67	−32.34	−32.86	−33.31	−33.61
2.75	−28.56	−28.98	−29.25	−29.42	−29.53	−30.09
3.00	−25.32	−25.11	−24.79	−24.40	−24.02	−22.15

6.3   The Non-isothermal Gray Medium

In this section we deal with an enclosure filled in with a gray emitting-absorbing medium. To calculate the coefficients $B(\varphi_i, \varphi_j)$ and $L(\varphi_i)$ of the linear system $B \cdot H_{\varepsilon,h(n)} = L$, line integrations need to be performed. The forcing term $F(p)$ of our integral equation is given by:

$$F(p) = \int_S (\varepsilon(r)e_b(r)\tau(r,p) + L(r,p))K(r,p)dS(r) .$$

The transmissivity $\tau(r,p)$ is defined as:

$$\tau(r,p) = \exp\left(-\int_{(r,p)} a(u)dL(u)\right) .$$

To illustrate the applicability of our algorithm in non-isothermal medium, or in furnaces filled up with a medium that exhibit considerable temperature gradients, we shall consider two examples, although somewhat artificial, of a short and long jet flame in a cylindrical furnace.

6.3.1   Radiation from Jet Flames

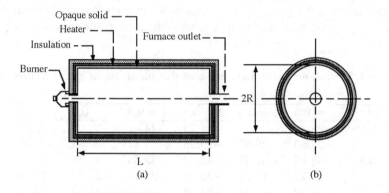

Figure 6.4. A cylindrical shape furnace

Figure 6.4 above shows the case considered. A cylindrical shape furnace of sizes $2R = 1m$ in diameter and $L = 3m$ long accommodates the flames. The furnace wall emissivity is assumed to be constant and equal to $\varepsilon = 0.7$. The furnace is axially fired and symmetrically cooled about the axis (Oz). The medium temperature rises from an inlet value T_o, passes through the adiabatic flame temperature T_a, and decreases continuously toward the bottom wall $(z = L)$. At each cross-section, the temperature distribution is assumed to possess a Gaussian profile. Therefore, we may represent the temperature distribution at every point $p(x, y, z)$ inside the furnace by an expression of the form:

$$T(p) = T_o(p) + (T_a - T_o(p))g(p)$$
$$T_o(p) = \frac{T_i}{2}\left(1 + \frac{1}{2R}\sqrt{x^2 + y^2}\right)$$
$$g(p) = \frac{L}{z}\exp\left(-m_o - \frac{1}{2}\left(\left(\frac{\ln(z/L) + m_o}{s}\right)^2 + \frac{x^2 + y^2}{(\sigma R)^2}\right)\right)$$

In this relation, the parameters m_o and $s > 0$ determine the position of the maximum temperature of the jet along the axis of symmetry (Oz); the parameter $\sigma > 0$ determines the jet width and the angle of spread of the jet along the axis of symmetry (Oz); and T_i is the temperature at the position $O(x = 0, y = 0, z = 0)$. It is further assumed that the above formula when applied at the points $p(x, y, z)$ such that $R = \sqrt{x^2 + y^2}$ and/or $z \in \{0, \ L\}$ gives the furnace wall temperature. There is no difference between gas and wall temperature at the furnace boundary surface. In attempting to arrive at representative temperature distributions observed in furnace practices, we assign the following values: $T_i = 1200°C$; $T_a = 1800°C$; $m_o = 0.750, 0.250$; $s = 1.000$; $\sigma = 0.375$.

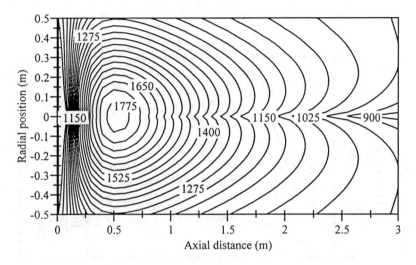

Figure 6.5. Temperature distribution for the short flame case

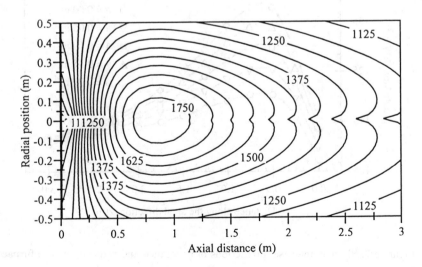

Figure 6.6. Temperature distribution for the long flame case

The profiles of temperature distribution for both flames are shown in Fig.6.5 and Fig. 6.6. The temperature rises from the interior front wall, passes through a maximum, and fall off rapidly as the combustion products fill the cross sections in the last meter of the furnace volume. The case $m_o = 0.750$ is designed to represent a short flame with high heat extraction (furnace exit temperature 900°C), and the case $m_o = 0.250$ a longer flame with moderate heat extraction (furnace exit temperature 1100°C).

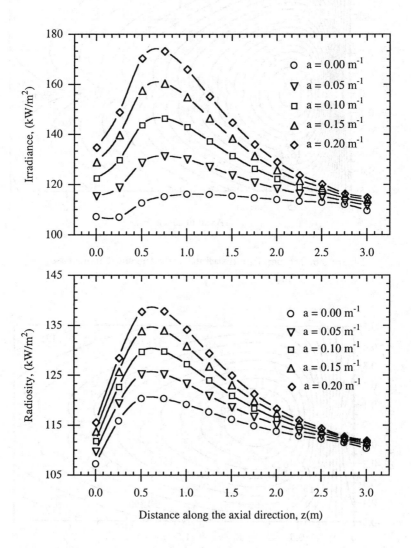

Figure 6.7. Short flame case; predictions of irradiance and radiosity at the furnace wall

Table 6.5. Short flame case; predictions of irradiance and radiosity distributions at the furnace side wall

	Irradiance, kW/m^2				
	absorption coefficient, $a(m^{-1})$				
$z(m)$	0.00	0.05	0.10	0.15	0.20
0.00	107.13	115.47	122.61	129.07	134.89
0.25	107.00	118.88	129.84	139.88	149.07
0.50	112.58	128.83	143.81	157.68	170.54
0.75	115.10	131.43	146.53	160.50	173.45
1.00	116.07	130.17	143.14	155.11	166.17
1.25	115.98	127.13	137.33	146.69	155.29
1.50	115.41	123.84	131.50	138.48	144.86
1.75	114.67	120.86	126.45	131.50	136.09
2.00	113.93	118.40	122.39	125.97	129.19
2.25	113.28	116.42	119.20	121.66	123.87
2.50	112.94	115.19	117.11	118.80	120.27
2.75	112.07	113.39	114.57	115.58	116.46
3.00	109.58	111.63	112.93	114.06	115.00

	Radiosity, kW/m^2				
	absorption coefficient, $a(m^{-1})$				
$z(m)$	0.00	0.05	0.10	0.15	0.20
0.00	107.31	109.81	111.95	113.89	115.64
0.25	115.91	119.48	122.76	125.77	128.53
0.50	120.40	125.27	129.77	133.93	137.79
0.75	120.40	125.29	129.82	134.01	137.90
1.00	119.17	123.40	127.29	130.88	134.20
1.25	117.65	121.00	124.06	126.87	129.45
1.50	116.20	118.73	121.03	123.12	125.04
1.75	114.93	116.78	118.46	119.98	121.35
2.00	113.85	115.19	116.38	117.46	118.42
2.25	112.95	113.89	114.73	115.47	116.13
2.50	112.28	112.96	113.53	114.04	114.48
2.75	111.55	111.95	112.30	112.61	112.87
3.00	110.42	111.03	111.42	111.76	112.05

For both flames we shall compute the irradiance and radiosity distributions at the heat sink for several constant values of absorption coefficient.

Figure 6.7 shows the distributions of irradiance and radiosity computed for the short flame. From this figure we observe that the incident heat fluxes, as well as the outgoing heat fluxes, at the furnace interior surface increase with the medium absorption coefficient. Their numerical values

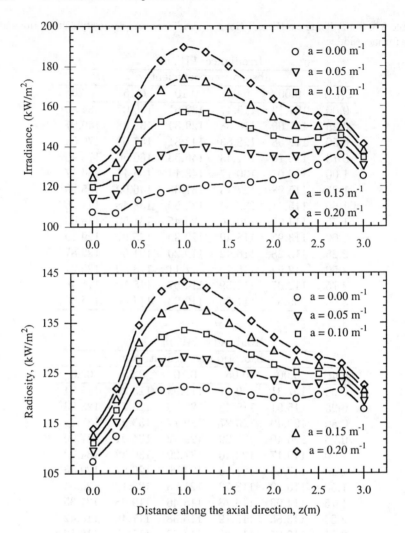

Figure 6.8. Long flame case; predictions of irradiance and radiosity at the furnace wall

are listed in Tables 6.5, for $r = 0.5m$. Both the irradiance and the radiosity at the side wall rise with the distance from the front wall, passe through a maximum, and fall off as the temperature decreases.

Some representative results for the long flame case are shown in Fig. 6.8 and their numerical values listed in Table 6.6. The position of the maxima for both the incident heat flux and the outgoing heat flux are located further downstream of the furnace since these are directly determined by the in-furnace temperature distribution.

Table 6.6. Long flame case; predictions of irradiance and radiosity distributions at the furnace side wall

Irradiance, kW/m^2

$z(m)$	absorption coefficient, $a(m^{-1})$				
	0.00	0.05	0.10	0.15	0.20
0.00	107.39	114.33	120.03	125.04	129.42
0.25	106.88	116.18	124.57	132.08	138.80
0.50	113.20	128.17	141.80	154.26	165.67
0.75	116.91	135.57	152.72	168.50	183.04
1.00	119.26	139.11	157.36	174.17	189.68
1.25	120.50	139.37	156.69	172.62	187.28
1.50	121.24	137.98	153.29	167.33	180.21
1.75	121.95	136.14	149.06	160.86	171.65
2.00	123.17	134.83	145.40	155.01	163.75
2.25	125.53	134.84	143.23	150.82	157.70
2.50	130.57	137.89	144.40	150.24	155.48
2.75	135.66	140.89	145.60	149.77	153.49
3.00	125.03	130.31	134.44	138.10	141.30

Radiosity, kW/m^2

$z(m)$	absorption coefficient, $a(m^{-1})$				
	0.00	0.05	0.10	0.15	0.20
0.00	107.39	109.47	111.18	112.68	113.99
0.25	112.43	115.22	117.74	119.99	122.01
0.50	118.96	123.45	127.54	131.28	134.70
0.75	121.60	127.20	132.34	137.07	141.44
1.00	122.28	128.23	133.71	138.75	143.40
1.25	121.97	127.63	132.83	137.61	142.00
1.50	121.29	126.31	130.90	135.11	138.97
1.75	120.55	124.81	128.69	132.23	135.46
2.00	120.02	123.52	126.69	129.57	132.19
2.25	119.90	122.69	125.21	127.49	129.55
2.50	120.67	122.87	124.82	126.58	128.15
2.75	121.54	123.11	124.52	125.78	126.89
3.00	117.78	119.36	120.60	121.70	122.66

The numerical simulations in this third exercise were also performed here with 254 nodal points, hence 504 triangular surface elements, placed over the furnace interior surface for each case. We have used per surface elements, 75 quadratures points for inner integrals and 4 quadrature points for outer integrals. Further we use the trapezoidal rule with equally space points, one centimeter spacing, for calculation of line integrals throughout the furnace volume. The results given here took approximately 8 minutes per run of a machine equipped with a Pentium II processor 400 MHz; 4 minutes for calculation of the matrix B of the linear system $B \cdot H_{\varepsilon,h(n)} = L$, 02 minutes for calculation of the vector L, and 02 minutes for solving the linear system $B \cdot H_{\varepsilon,h(n)} = L$ of size 254×254 using Cholesky method.

Up to this point we have concerned ourselves exclusively and implicitly with radiation at a single wavelength or in a gray medium. We must now go on to consider radiation in spectral bands.

6.4   Non Gray Medium; Band Approximation

Most industrial combustion chambers are gas-filled enclosures. The radiative properties of the media are always wavelength dependent. As explained in Chap. 2, gases are transparent to radiation in some regions of the spectrum (windows) while within other spectral intervals (bands) they absorb or emit the radiant energy. The nature of this interaction is determined by molecular spectra and the exact description of this phenomena involves quantum mechanics. A discussion on the spectral characteristics of gaseous media is provided in the succeeding chapter. Each band comprises large number of spectral lines. The shape of these lines as well as their overlapping depend on several factors.

Numerically, we can account for the band radiation by assuming that the radiative properties are independent of wavelength in each spectral band $\Delta\eta$. In this case the analysis developed in the foregoing sections remains valid as long as it is applied to each wavelength interval $\Delta\eta$. The forcing term $F(p, \Delta\lambda)$ over the wavelength interval $\Delta\eta$ is defined as:

$$F(p, \Delta\lambda) = \int_S (\varepsilon(r, \Delta\lambda)e_b(r, \Delta\lambda)\tau(r, p, \Delta\lambda) + L(r, p, \Delta\lambda))\, K(r, p)dS(r) \ .$$

For an isothermal medium of temperature T^m, blackbody emissive power $E_b(T^m, \Delta\lambda)$, within the enclosure, the line integral $L(r, p, \Delta\lambda)$ over the wavelength interval $\Delta\eta$ is given as:

$$L(r, p, \Delta\lambda) = E_b(T^m, \Delta\lambda)\,(1 - \tau(r, p, \Delta\lambda)) \ ,$$

whereas for a non-isothermal medium filling up the enclosure volume, it is given as:

$$L(r, p, \Delta\lambda) = \int_{(r,p)} a(u, \Delta\lambda)e_b(u, \Delta\lambda)\tau(u, p, \Delta\lambda)dL(u) \ .$$

The transmissivity in the same wavelength interval $\Delta\eta$ is defined to be:

$$\tau(r, p, \Delta\lambda) = \frac{1}{|\Delta\lambda|} \int_{\Delta\lambda} \exp\left(-\int_{(r,p)} a(u, \lambda)dL(u)\right) d\lambda \ .$$

The unknowns of the linear system are the values $H_{\varepsilon,h(n)}(p, \Delta\lambda)$ defined as:

$$H_{\varepsilon,h(n)}(p, \Delta\lambda) = \int_{\Delta\lambda} H_{\varepsilon,h(n)}(p, \lambda)d\lambda \ .$$

The approximated total heat $-$ux, including all wavelengths, $q_{h(n)}(p)$ at any point p of the enclosure boundary is calculated by summing up the contribution of the $N_{\Delta\lambda}$ different spectral bands. Accordingly,

$$q_{h(n)}(p) = \sum_{i=1}^{N_{\Delta\lambda}} \varepsilon(p, (\Delta\lambda)_i) \cdot \left(e_b(p, (\Delta\lambda)_i) - \frac{H_{\varepsilon,h(n)}(p, (\Delta\lambda)_i)}{\sqrt{1 - \varepsilon(p, (\Delta\lambda)_i)}}\right),$$

where $(\Delta\lambda)_i = (\lambda_i, \lambda_{i+1})$. The total net radiative heat source is determined in a similar manner.

We shall discuss in the next chapter an example showing the ability of our algorithm to handle a non gray medium.

7
Spectral Properties of Gases

This chapter dwells on the spectral radiative properties of gaseous media. The spectral transmissivity $\tau(r, p, \Delta\lambda)$, averaged over the band interval $\Delta\lambda$, and an appropriately chosen absorption coefficient $a(p, \Delta\lambda)$ are needed to calculate the coefficients $B(\varphi_i, \varphi_j)$ and $L(\varphi_i)$ of the linear system $B \cdot H_{\varepsilon,h(n)} = L$, and the net radiative heat source $q_V(p)$. Here, the term $\Delta\lambda$ refers to the wavelength interval over which the spectral absorption coefficient is nonzero. To facilitate consistency with the literature on spectroscopy, we use in the present chapter a notation slightly deviating from that adopted in Chap. 2. Instead of the wavelength variable λ we shall use the wavenumber η whenever the meaning is clear.

The transmissivity is a function of the path length, temperature, pressure and emitter density. In furnaces, the pressure variations are negligible compared to the temperature variations. Consequently, we will assume in this chapter that the non-homogeneity along a path length (r, p), when it occurs, is due to the temperature and gas composition variations only. The closed form expression for the transmissivity $\tau(r, p, \Delta\lambda)$ presented in this chapter originates from the work of Edwards and Menard [118]-[120], and the developments of Felske and Tien [121]-[122]. It is not claimed that the information given in this chapter is complete in any sense and we do not intend to enter into an exhaustive discussion of spectral radiative properties of gases. For further details on spectral bands radiation calculations we refer the reader to the literature [123]-[133]. Recently an extensive study on several radiative properties models and their coupling with numerical methods of radiative heat transfer has also been carried out at the IFRF [88, 89].

7.1   Principle of Infrared-Radiation in Gases

A gas molecule emits radiation as the result of electronic, vibrational, and rotational transitions from excited energy levels to lower energy levels.

7.1.1   Electronic Transitions

Electronic transitions are generally divided into three types: bound-bound transitions, bound-free transitions and free-free transitions, as illustrated in Fig. 7.1. In this figure, the energy E_I is the ionization potential; that is, the energy required to produce ionization from the ground state.

In bound-bound transitions, a photon interacting with a molecule causes a change from one bound level to another without ionization. This interaction produces a line spectra.

In bound-free transitions, a photon interacting with a bound electron-atom system produces a system which has too much energy for the electron to ionize the gas species.

In free-free transitions, a free electron can either emit a photon without losing all of its kinetic energy or it can absorb a photon and its kinetic energy increases.

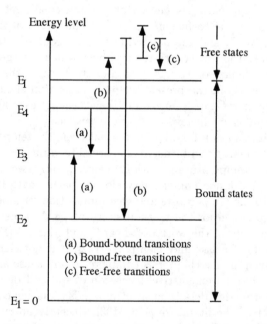

Figure 7.1. Schematic diagram of energy states and transitions for atoms, ions or electrons

Energy level

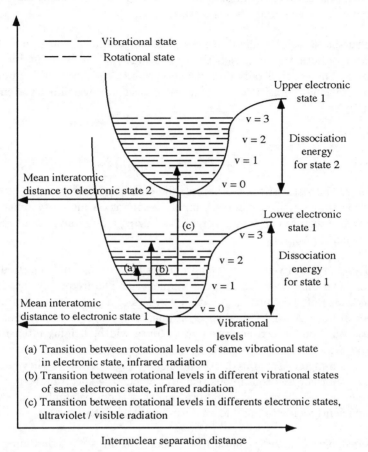

Figure 7.2. Schematic diagram of energy states and transitions for diatomic molecules

7.1.2 Vibrational and Rotational Transitions

Vibrational transitions are always accompanied with rotational transitions. They can occur only in those molecules which have an oscillating dipole associated with the vibrational or rotational energy state. Such infrared-radiating gases include H_2O, CO_2, CO and NO.

Figure 7.2 illustrates the large number of possible energy transitions that can produce an array of vibrational-rotational lines spectra. This figure shows the potential energy for a diatomic molecule as a function of the separation distance between its two atoms. The two curves are for different electronic energy states where the electron may be shared by the two atoms.

Transitions between rotational levels of a given vibrational state give a series of closely spaced lines spectra.

The above basic atomic processes contribute to the absorption and emission of radiation as it passes through a material, or as it interacts with matter. The emitted radiation, which is due to the electronic, vibrational and rotational transitions, is distributed over a well-defined wavenumber region $\Delta\eta$.

7.1.3 Spectral Lines of Emitted Radiation

If we refer to statistical mechanics, the emitting-absorbing gas molecules may be thought of as complex, highly resonant harmonic systems. Their internal structure can be interpreted as having a large number of resonant frequencies [123].

To satisfy the fixed energy requirements for a change in excitation, transitions can only occur at discrete frequencies. The frequency distribution (spectrum) of the emitted radiation consists of a large number of distinct spectral lines. An individual spectral line, however, is not truly monochromatic but can be characterized by a finite width, height, intensity, and shape. At sufficiently low pressures the spectral lines can be considered as to be completely separated. At sufficiently high pressures they merge to form a more or less continuous region in which the spectral absorption coefficient varies slightly. We shall assume that the wavenumber interval $\Delta\eta$ is sufficiently wide to include an entire spectral line.

When the emitted radiation is observed from a gas containing many molecules including polyatomic molecules such as carbon dioxide and water vapor, the molecules moving toward the observer will appear to have slightly increased frequency of emission and those moving away from the observer will seem to have a lower frequency. The random motion of polyatomic molecules, or of simpler molecules in high temperatures, in a gas often show a random distribution of spectral line positions. This gives rise to the consideration of probability distributions for describing the behavior of spectral lines.

The shape of a typical spectral line is illustrated in Fig. 7.3. In describing the behavior of spectral lines, the parameter which is frequently regarded as a random variable due to random motion of gas molecules is the line strength S. The line strength $S(T^m(p), \eta_c)$ of a particular spectral line centered at the wavenumber η_c, at a temperature $T^m(p)$, measures the total absorption when the optical depth is small enough to eliminate self-absorption. It is defined as the wavenumber integral of the spectral contour

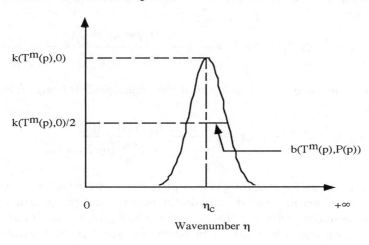

Figure 7.3. Shape of a spectral line

$k(T^m(p), \eta - \eta_c)$ of the spectral line:

$$S(T^m(p), \eta_c) = \int_{\infty}^{+\infty} k(T^m(p), \eta - \eta_c) d(\eta - \eta_c) \, .$$

In a random spectrum of Lorentz lines of equal widths, the probability distribution of lines strength may be described using the Mayer–Goody statistical model [124]. Precise calculations of the radiative properties of emitted radiation should be performed for each spectral line of the spectrum using such probability distributions. In this case the absorption coefficient $a(p, \eta)$ at the wavenumber η produced by the spectral line, which is centered at η_c, is defined to be:

$$a(p, \eta) = S(T^m(p), \eta_c) s(\eta - \eta_c, b(T^m(p), P(p))) \rho(p) \, ,$$

with $s(\eta - \eta_c, b(T^m(p), P(p)))$ being the line-shape parameter at the temperature $T^m(p)$ and pressure $P(p)$, $b(T^m(p), P(p))$ being a measure of the half-width at half-height of the spectral line contour, and $\rho(p)$ being the density of the absorbing gas component at the point p. We also have

$$\int_{\infty}^{+\infty} s(\eta - \eta_c, b(T^m(p), P(p))) d\eta = 1 \, .$$

For Lorentz lines, the dispersion distribution or line-shape parameter is given as:

$$s(\eta - \eta_c, b(T^m(p), P(p))) = \frac{b(T^m(p), P(p))}{\pi \cdot \left((\eta - \eta_c)^2 + (b(T^m(p), P(p)))^2\right)} .$$

We may therefore rewrite the spectral absorption coefficient $a(p, \eta)$ in the form:

$$a(p, \eta) = S(T^m(p), \eta_c) \frac{b(T^m(p), P(p)) \cdot \rho(p)}{\pi \cdot \left((\eta - \eta_c)^2 + (b(T^m(p), P(p)))^2\right)} . \qquad (7.1)$$

It is worth noting that the gas absorption coefficient, which is in fact an emission coefficient, depends only on the temperature, the pressure, and the medium composition at the local point p being considered. It does not on the path length traverses by radiation. The spectral transmissivity is defined to be:

$$\tau(r, p, \eta) = \exp\left(-\int_{(r,p)} a(u, \eta) dL(u)\right) . \qquad (7.2)$$

Although calculations for each spectral line are feasible, the computer time required for determining the radiative properties over the entire spectral region is too excessive for engineering practices. However, in practice, the emission lines do not occur in isolation but are spaced so closely together that the broadened lines overlap to a considerable extent. To avoid the problem of formulating the radiative properties as complex functions of wavenumber and to reduce their computational time, hypothetical models of simplified mathematical structure, called band models, are introduced. A band is defined as a series of spectral lines that produce nonzero absorption coefficients at wavenumbers in the vicinity of the specified band center. It provides a fair approximation of the spectrum of emitted radiation.

In modeling the radiative behavior of spectral bands we may employ two types of band models. The first type, called narrow-band model, uses the characterizations of individual spectral line shape, width, and spacing to derive the band properties. The second type, called wide-band model, provides the correlations of the band characteristics valid over the entire wavenumber region of the band.

7.2   Properties of an Isothermal Gas Species

In this section we shall concern ourselves with determining the spectral transmissivity $\tau(r, p, \Delta\eta)$ over the band interval $\Delta\eta$ for an isothermal gas species.

We consider a group of lines with a random line position selected from a Lorentz dispersion distribution and a random line intensity from the exponential distribution

$$S \longmapsto \frac{1}{\bar{S}} \cdot \exp\left(-\frac{S}{\bar{S}}\right) ,$$

with $\bar{S}$ being the lines mean strength.

We assume that there are no correlation between the line positions and the line intensities. Then, the Mayer–Goody statistical model predicts the mathematical expectation of the spectral transmissivity of a narrow band $\delta\eta$ for an isothermal gas at the temperature T^m and total pressure P to be:

$$\tau(r,p,\delta\eta) = \exp\left(-\frac{\bar{S}(T^m,\delta\eta)}{d}X(r,p)\sqrt{1 + \frac{\bar{S}(T^m,\delta\eta)}{d}\frac{X(r,p)}{\beta(T^m,P_e)}}\right) \quad (7.3)$$

with d being the mean spectral line spacing, $\bar{S}(T^m,\delta\eta)/d$ the mean spectral line intensity to spacing ratio, $\beta(T^m,P_e) = 4b(T^m,P_e)/d$ the pressure broadening parameter, P_e the effective broadening pressure which depends on P, and $X(r,p) = \rho \cdot \|r-p\|_{R^3}$ the mass path length of the absorbing gas component.

The mean spectral line to spacing ratio $\bar{S}(T^m,\delta\eta)/d$ and the parameter $\beta(T^m,P_e)$ can be considered as the fundamental properties of a gas, as far as radiation is concerned. It is the manner in which these parameters vary over the spectrum that determines the spectral bands. The parameter $\beta(T^m,P_e)$ is important as it specifies the effect of line structure.

For large $\beta(T^m,P_e)$, the lines become broad compare to their spacing and the structure of a single line is spread over the band. This yields a narrow band transmissivity

$$\tau(r,p,\delta\eta) = \exp\left(-\frac{\bar{S}(T^m,\delta\eta)}{d}X(r,p)\right) .$$

For very small $\beta(T^m,P_e)$, the lines are thin and spaced well apart. Each single line acts independently and for strong non-overlapping lines this yields a narrow band transmissivity

$$\tau(r,p,\delta\eta) = \exp\left(-\sqrt{\frac{\bar{S}(T^m,\delta\eta)}{d}X(r,p)\beta(T^m,P_e)}\right) .$$

In the above relations $\left(\bar{S}(T^m,\delta\eta)/d\right) \cdot \rho$ is an effective absorption coefficient of the narrow band.

To specify the wavenumber variations of the parameters $\bar{S}(T^m,\delta\eta)/d$ and $\beta(T^m,P_e)$ over the band interval, we may utilize Edwards' wide band model. This model assumes the effective broadening pressure P_e to be defined in the form:

$$P_e = (P + x \cdot (b-1))^n \, ,$$

where P is the total pressure of radiating and non-radiating gas, and x is the mole fraction of the radiating gas. Here, b is the self-broadening to foreign-gas broadening ratio and n is the pressure broadening exponent.

Edwards' model also assumes the mean spectral line intensity to spacing ratio $\bar{S}(T^m, \delta\eta)/d$ to vary only slowly with wavenumber across the entire wide band interval of width $\Delta\eta$ and not from spectral line to spectral line [120]-[121]. This variation is taken to be exponential using the form:

$$\frac{\bar{S}(T^m, \delta\eta)}{d} = \frac{\alpha(T^m)}{\omega(T^m)} \exp\left(\frac{-|\eta - \eta_o|}{\omega(T^m)}\right) \, , \qquad (7.4)$$

for band intervals with upper or lower wavenumber heads at η_o.

In the expression above, $\alpha(T^m)$ is the integrated band intensity and $\omega(T^m) = \omega_o \cdot (T^m/T_o)^{1/2}$ is the band width parameter with ω_o being the band width parameter at the reference temperature T_o. The integrated band intensity $\alpha(T^m)$ and the pressure broadening parameter $\beta(T^m, P_e)$ are obtained using statistical mechanics considerations. Their expressions are given by the relations:

$$\alpha(T^m) = \alpha_o \frac{1 - \exp\left(-\sum_{k=1}^{m} u_k(T^m)\delta_k\right)}{1 - \exp\left(-\sum_{k=1}^{m} u_k(T_o)\delta_k\right)} \frac{\Psi(T^m)}{\Psi(T_o)}$$

$$\beta(T^m, P_e) = \beta_o \sqrt{\frac{T_o}{T^m}} \frac{\Phi(T^m)}{\Phi(T_o)} P_e$$

$$u_k(T^m) = C_2 \frac{\eta_k}{T^m} \, .$$

The functions $\Psi(T^m)$ and $\Phi(T^m)$ are defined to be:

$$\Psi(T^m) = \frac{\prod_{k=1}^{M} \sum_{v_k=v_{o,k}}^{+\infty} \left(\frac{(v_k+g_k+|\delta_k|-1)!}{(g_k-1)!v_k!} \exp\left(-u_k(T^m)v_k\right)\right)}{\prod_{k=1}^{M} \sum_{v_k=0}^{+\infty} \left(\frac{(v_k+g_k-1)!}{(g_k-1)!v_k!} \exp\left(-u_k(T^m)v_k\right)\right)} \, ,$$

$$\Phi(T^m) = \frac{\left(\prod_{k=1}^{M} \sum_{v_k=v_{o,k}}^{+\infty} \sqrt{\frac{(v_k+g_k+|\delta_k|-1)!}{(g_k-1)!v_k!}} \exp\left(-u_k(T^m)v_k\right)\right)^2}{\prod_{k=1}^{M} \sum_{v_k=0}^{+\infty} \left(\frac{(v_k+g_k-1)!}{(g_k-1)!v_k!} \exp\left(-u_k(T^m)v_k\right)\right)} \, ,$$

with $v_{o,k} = 0$ if $\delta_k \geq 0$, $v_{o,k} = |\delta_k|$ and otherwise.

The parameters required to obtaining $\alpha(T^m)$, $\beta(T^m, P_e)$, and $\omega(T^m)$ for the infrared absorption band of the commonly encountered gases in industrial furnaces are listed in Table 7.1 through Table 7.4 in the next section. An optimized procedure for calculating the above parameters can be found in reference [134].

The parameter $\bar{S}(T^m, \delta\eta)/d$ varies only slowly with wavenumber across the entire band interval and does not vary from line to line. Therefore, we may approximate the spectral transmissivity $\tau(r, p, \Delta\eta)$ within the band interval $\eta_l \leq \eta \leq \eta_u$ of width $|\Delta\eta| = \eta_u - \eta_l$ by an average value of the population of narrow band transmissivities $\tau(r, p, \delta\eta)$. This average spectral transmissivity, also denoted by $\tau(r, p, \Delta\eta)$ for convenience, can be defined as:

$$\tau(r, p, \Delta\eta) \equiv 1 - \frac{1}{|\Delta\eta|} \int_{\Delta\eta} (1 - \tau(r, p, \delta\eta)) \, d\eta \ .$$

The spectral absorption coefficient is nonzero in the wavenumber interval $\Delta\eta$ but takes zero (negligible) value outside. If we define the function $g(u, \beta, x)$ by the relation:

$$g(u, \beta, x) = 1 - \exp\left(-u \cdot e^{-x} \cdot \sqrt{1 + \frac{u}{\beta} e^{-x}}\right)$$

then, it follows that:

$$\tau(r, p, \Delta\eta) = 1 - \frac{1}{|\Delta\eta|} \int_0^{+\infty} g(u(r, p, T^m), \beta(T^m, P_e), \frac{|\eta - \eta_o|}{\omega(T^m)}) d(\eta - \eta_o).$$

In this relation the integrated optical depth parameter $u(r, p, T^m)$ for the gas component being considered is defined as:

$$u(r, p, T^m) = \frac{\alpha(T^m)}{\omega(T^m)} X(r, p) \ .$$

If we introduce the variable of integration [121]:

$$\xi = \frac{\zeta^2 - 1}{\zeta} \ ,$$

where

$$\zeta = \sqrt{1 + \frac{u(r, p, T^m)}{\beta(T^m, P_e)} \cdot \exp\left(\frac{-|\eta - \eta_o|}{\omega(T^m)}\right)}$$

then, average spectral transmissivity may be calculated explicitly. It is given by the relation:

$$\tau(r, p, \Delta\eta) = 1 - \frac{\omega(T^m)}{|\Delta\eta|} \left(F(r, p, T^m, P_e) + G(r, p, T^m, P_e)\right) \ , \qquad (7.5)$$

with

$$F(r, p, T^m, P_e) = E_1(U(r, p, T^m, P_e)) + \ln(U(r, p, T^m, P_e)) + \gamma_{Euler} \ . \qquad (7.6)$$

The function $E_1(\cdot)$ represents the first order exponential integral, γ_{Euler} is the Euler–Mascheroni constant, and the function $U(r, p, T^m, P_e)$ is defined by the relation:

$$U(r, p, T^m, P_e) = \sqrt{\frac{1}{u(r, p, T^m)} \left(\frac{1}{\beta(T^m, P_e)} + \frac{1}{u(r, p, T^m)} \right)} . \qquad (7.7)$$

The integral $G(r, p, T^m, P_e)$ is given as:

$$G(r, p, T^m, P_e) = \int_0^{\frac{U(r,p,T^m,P_e)}{\beta(T^m,P_e)}} \frac{1 - \exp\left(-\beta(T^m, P_e) \cdot \xi\right)}{\sqrt{\xi^2 + 4}} d\xi .$$

By using the inverse hyperbolic sine function, we may also express the function $G(r, p, T^m, P_e)$ in the form:

$$\frac{G(r, p, T^m, P_e)}{H(r, p, T^m, P_e)} = \int_0^{\frac{U(r,p,T^m,P_e)}{\beta(T^m,P_e)}} \frac{\exp\left(-\beta(T^m, P_e) \cdot \xi\right)}{\sqrt{\xi^2 + 4}} d\xi , \qquad (7.8)$$

$$H(r, p, T^m, P_e) = \text{arcsinh}\left(\frac{U(r, p, T^m, P_e)}{2\beta(T^m, P_e)} \right). \qquad (7.9)$$

The integral term appearing in the relation hereabove cannot be simplified into a closed form. It will be evaluated exactly by numerical integration.

7.2.1 Exponential Wide-Band Parameters

For the H_2O molecule, the self-broadening to foreign-gas ratio is given as:

$$b(T) = 8.6 \cdot \sqrt{\frac{T_o}{T}} + 0.5 .$$

Triatomic molecules are characterized by three modes of vibration: symmetric stretching, bending, and asymmetric stretching. The fundamental vibration wavenumber are listed as η_1, η_2, and η_3 corresponding to each of the three fundamental mode of vibration. The statistical weight factors g_k, $k = 1, 2, 3$ indicate the number of ways that a particular vibration mode can occur. The empirical fudge factor n represents the pressure broadening exponent.

The CO molecule has one ($M = 1$) fundamental vibration mode. The fundamental (i.e. a band with a single non-zero positive value of δ_1 equal to unity) oscillator wavenumber for stretching the CO molecule η_1 is equal to 2143 cm^{-1}, the corresponding wavelength is $\lambda_1 = 4.67\mu m$ and the associated statistical degeneracy g_1 is equal to 1. The first overtone (i.e. a

Table 7.1. Exponential wide band parameters for H_2O [68],[118],[119]

Fundamental vibration mode $(M = 3)$	Fundamental wavenumber, cm^{-1}	Statistical degeneracy
Symmetric stretching	$\eta_1 = 3652$	$g_1 = 1$
Bending	$\eta_2 = 1596$	$g_2 = 1$
Asymmetric stretching	$\eta_3 = 3756$	$g_3 = 1$

Bands	Vibrational transitions	Pressure parameter		Band center	Absorption parameters		
μm	$\delta_1,\delta_2,\delta_3$	$T_o = 100K$		η_c	α_o	β_o	ω_o
		n	b				
> 10	$0,0,0$	1.00	$b(T)$	140	44205	0.14311	69.30
6.30	$0,1,0$	1.00	$b(T)$	1600	41.20	0.09427	56.40
2.70	$0,2,0$	1.00	$b(T)$	3760	0.19	0.13219	60.00
	$1,0,0$				2.30		
	$0,0,1$				22.40		
1.87	$0,1,1$	1.00	$b(T)$	5350	3.00	0.08169	43.10
1.38	$1,0,1$	1.00	$b(T)$	7250	2.50	0.11628	32.00

Table 7.2. Exponential wide band parameters for CO_2 [68],[118]

Fundamental vibration mode $(M = 3)$	Fundamental wavenumber, $1/cm$	Statistical degeneracy
Symmetric stretching	$\eta_1 = 1351$	$g_1 = 1$
Bending	$\eta_2 = 667$	$g_2 = 2$
Asymmetric stretching	$\eta_3 = 2396$	$g_3 = 1$

Bands	Vibrational transitions	Pressure parameter		Band center	Absorption parameters		
μm	$\delta_1,\delta_2,\delta_3$	$T_o = 100K$		η_c	α_o	β_o	ω_o
		n	b				
15.00	$0,1,0$	0.70	1.30	667	19.00	0.06157	12.70
10.00	$-1,0,1$	0.80	1.30	960	$2.47e-9$	0.04017	13.40
9.40	$0,-2,0$	0.80	1.30	1060	$2.48e-9$	0.11888	10.10
4.30	$0,0,1$	0.80	1.30	2410	110.00	0.24723	11.20
2.70	$1,0,1$	0.65	1.30	3660	4.00	0.13341	23.50
2.00	$2,0,1$	0.65	1.30	5200	$6.60e-2$	0.39305	34.50

Table 7.3. Exponential wide band parameters for CO [68]-[118]

Bands	Vibrational transitions	Pressure parameter		Band center	Absorption parameters		
μm	δ_1	$T_o = 100K$		η_c	α_o	β_o	ω_o
		n	b				
4.67	1	0.80	1.10	2143	20.90	0.07506	25.50
2.35	2	0.80	1.00	4260	0.14	0.16758	20.00

Table 7.4. Exponential wide band parameters for NO [68]-[118]

Bands	Vibrational transitions	Pressure parameter		Band center	Absorption parameters		
μm	δ_1	$T_o = 100K$		η_c	α_o	β_o	ω_o
		n	b				
5.34	1	0.65	1.00	1876	9.00	0.18050	20.00

band with a single non-zero positive value of δ_1 greater than unity) occurs at $4260\ cm^{-1}$ or $2.35\ \mu m$.

The NO molecule has also one $(M = 1)$ fundamental vibration mode. The fundamental oscillator wavenumber for stretching the molecule η_1 is equal to $1876\ cm^{-1}$, and the associated degeneracy g_1 is equal to 1. The exponential wide band parameters for this molecule are given in Table 7.4.

7.2.2 Average Spectral Transmissivity

To evaluate the average spectral transmissivity $\tau(r, p, \Delta\eta)$ from the above considerations, the spectral interval $\Delta\eta$ over which the spectral absorption coefficient is nonzero has to be specified. The product

$$\omega(T^m) \cdot (F(r, p, T^m, P_e) + G(r, p, T^m, P_e))$$

appearing in (7.5) is often referred to as the total band absorptance or the effective band width and is denoted by $A(r, p, T^m, P_e)$. To estimate the average spectral transmissivity, the following relation is used [68]-[118]:

$$\tau(r, p, \Delta\eta) = \frac{u(r, p, T^m)}{A(r, p, T^m, P_e)} \frac{\partial A(r, p, T^m, P_e)}{\partial u(r, p, T^m)}.$$

For, applying Leibniz's Theorem for differentiation of an integral to the total band absorptance, the average spectral transmissivity is evaluated as:

$$\tau(r, p, \Delta\eta) = \frac{(U(r, p, T^m, P_e))^3}{2u(r, p, T^m)} \frac{B(r, p, T^m, P_e)C(r, p, T^m, P_e)}{F(r, p, T^m, P_e) + G(r, p, T^m, P_e)}, \quad (7.10)$$

where

$$B(r, p, T^m, P_e) = \frac{2}{u(r, p, T^m)} + \frac{1}{\beta(T^m, P_e)},$$

and

$$C(r, p, T^m, P_e) = \frac{C_1(r, p, T^m, P_e)}{U(r, p, T^m, P_e)} + \frac{C_1(r, p, T^m, P_e)}{C_2(r, p, T^m, P_e)}$$

$$C_1(r, p, T^m, P_e) = 1 - \exp(-U(r, p, T^m, P_e))$$

$$C_2(r, p, T^m, P_e) = \sqrt{(U(r, p, T^m, P_e))^2 + 4(\beta(T^m, P_e))^2}$$

The average spectral transmissivity $\tau(r, p, \Delta\eta)$ does not obey Bourguer's law.

Here, we merely point out that an average coefficient of absorption $a(p, \Delta\eta)$ defined to be

$$a(p, \Delta\eta) \equiv \frac{1}{|\Delta\eta|} \int_{\Delta\eta} a(p, \delta\eta) d\eta$$

for the spectral interval $\Delta\eta$ will have no physical meaning, since

$$\tau(r, p, \Delta\eta) \neq \exp\left(-\int_{\Delta\eta} a(u, \Delta\eta) dl(u)\right) .$$

Furthermore, by defining such quantity, we cannot take account of spectral correlations related to the line structure of the spectrum.

7.2.3 Spectral Band Limits

Having determine the average spectral transmissivity, (7.10), the spectral band width is obtained from (7.5). The result reads:

$$|\Delta\eta| = \frac{2u(r, p, T^m)\omega(T^m)\left(F(r, p, T^m, P_e) + G(r, p, T^m, P)\right)}{2u(r, p, T^m)\left(F(r, p, T^m, P_e) + G(r, p, T^m, P_e)\right) - D(r, p, T^m, P_e)}$$

where

$$D(r, p, T^m, P_e) = B(r, p, T^m, P_e)C(r, p, T^m, P_e)\left(U(r, p, T^m, P_e)\right)^3 .$$

The spectral band limits η_l and η_u for symmetrical gas band, with center at the wavenumber η_c, are given as:

$$\eta_l = \eta_c - \frac{1}{2}|\Delta\eta| , \quad \eta_u = \eta_c + \frac{1}{2}|\Delta\eta| .$$

At fixed pressure and temperature, the width $|\Delta\eta|$ of a spectral interval increases with mass path length. Therefore, it is not possible to set fix spectral band limits for use in the radiative heat exchange calculations. Nevertheless, the band centers (or one band limit) for the radiating gases encountered in furnaces are usually known. We may henceforth use the largest mass path length in the enclosure geometry being studied to subdivide the entire spectrum into fixed finite spectral intervals. As such the spectral intervals obtained include the widest band encountered. This subdivision of the entire spectrum is arbitrary but prove useful for engineering calculations.

7.3   Properties of a Non-isothermal Gas Species

This section is concerned with evaluating the average spectral transmissivity $\tau(r, p, \Delta\eta)$ for a non-isothermal gas medium. In the foregoing section we have presented the relations required to obtaining the spectral transmissivity $\tau(r, p, \Delta\eta)$ for an isothermal gas species. In industrial furnaces, however, the temperature varies substantially throughout the medium introducing additional complexities in determining the spectral transmissivity.

The basis for engineering treatment of non-isothermal gases is the Curtis–Godson method [123]. The Curtis–Godson method relates the spectral transmissivity $\tau(r, p, \Delta\eta)$ in a non-isothermal gas to the spectral transmissivity of an equivalent isothermal gas. The relation between the non-isothermal and the isothermal gas is carried out by assigning an equivalent amount of isothermal absorbing material to act in place of the non-isothermal gas. Under the assumption that the transmissivity of the non-uniform gas is equal to the spectral transmissivity of the uniform gas in the limit of small and large values of mass path length and pressure broadening parameter of the absorbing gas component, the following scaling parameters are obtained [120]-[122]:

$$\tilde{X}(r, p) = \int_{(r,p)} \rho(u)dL(u)$$

$$\tilde{\alpha}(T^m) = \frac{\int_{(r,p)} \alpha(T(u))\rho(u)dL(u)}{\tilde{X}(r, p)}$$

$$\tilde{\omega}(T^m) = \frac{\int_{(r,p)} \omega(T(u))\alpha(T(u))\rho(u)dL(u)}{\tilde{\alpha}(T^m)\tilde{X}(r, p)}$$

$$\tilde{\beta}(T^m, P_e) = \frac{\int_{(r,p)} \beta(T(u), P_e)\omega(T(u))\alpha(T(u))\rho(u)dL(u)}{\tilde{\omega}(T^m)\tilde{\alpha}(T^m)\tilde{X}(r, p)}$$

In these relations the term T^m represents the temperature of the isothermal medium. The average spectral transmissivity $\tau(r, p, \Delta\eta)$ is obtained upon substitution of the above scaling relations into (7.5) through to and including (7.10).

Although an average value $\tau(r, p, \Delta\eta)$ of the transmissivity is readily obtained from the algorithm above we are unable to perform radiative heat transfer calculations. This is due to the fact that to evaluate the line integral $L(r, p, \Delta\eta)$ for enclosures filled in with a non-isothermal gas mixture, the value of $a(p, \Delta\eta)$ for the spectral interval $\Delta\eta$ is needed. The absorption (emission) coefficient $a(p, \Delta\eta)$ determines the exponential attenuation of the intensity of incident radiation within the spectral interval $\Delta\eta$. It is a function of the state variables only. A general theory for determining an average gas absorption coefficient $a(p, \Delta\eta)$, that depends only on the local properties at the point p, would extend far beyond the scope of this book.

To obtain a tractable procedure for calculations of the line integrals $L(r, p, \Delta\eta)$ we observe first that the positive-valued functions that represent the medium spectral absorption coefficient $a(p, \eta)$, the spectral blackbody emissive power $e_b(T^m(u), \eta)$ and the spectral transmissivity $\tau(r, p, \eta)$ are continuous and bounded. Hence, from the mean value theorem, there exists a point u_o on the line of sight (r, p) such that the line integral $L(r, p, \Delta\eta)$ may be expressed in the form:

$$L(r, p, \Delta\eta) = e_b(T^m(u_o), \Delta\eta)\,(1 - \tau(r, p, \Delta\eta)) \ .$$

Determining the point u_o, however, is a difficult issue. Now, we realize that in determining the average spectral tranmissivity $\tau(r, p, \Delta\eta)$ we have assigned an equivalent amount of isothermal absorbing-emitting material to act in place of the non-isothermal gas component. Thus, our problem is reduced to the determination of the temperature T^m of the equivalent isothermal medium and we should have:

$$L(r, p, \Delta\eta) = e_b(T^m, \Delta\eta)\,(1 - \tau(r, p, \Delta\eta)) \ ,$$

which is independent of the choice of the point u_o.

7.4   Medium Containing Several Gas Species

In extending the procedure for determining the spectral radiative properties in medium containing several absorbing and emitting gases which may spectrally overlap (situation where adjacent bands of the same species; e.g., $9.4\mu m$ and $10.4\mu m$ CO_2 bands; or spectral intervals of different species, e.g. $2.7\mu m$ CO_2 and H_2O bands, overlap spectrally), we shall assume that the spectral transmissivities are uncorrelated since the spectral lines exhibit random behavior [125]. Then, the average transmissivities are multiplictives. This is true for strictly monochromatic radiation[1] where the spectral absorption coefficients are additives. Namely,

$$\tau(r, p, \eta) = \prod_k \tau_k(r, p, \eta) \ , \quad a(p, \eta) = \sum_k a_k(p, \eta)$$

with $\tau_k(r, p, \eta)$ and $a(p, \eta)$ being the spectral transmissivity and the spectral aborption coefficient of the individual chemical species k contributing to the gas radiation.

---

[1]Radiative heat transfer of a single wavenumber is also called monochromatic radiation since, over the visible range, the human eye perceives electromagnetic waves to have the colors of the rainbow.

For isothermal media, the transmissivities $\tau_k(r, p, \Delta\eta)$ are calculated by using (7.10). For non-isothermal media, the scaling parameters described above are used.

7.5   Band Radiation in a Well-Stirred Chamber

The objectives of the study in this section are: firstly, to demonstrate the ability of our algorithm to incorporate spectral calculations; secondly, to examine the effects of spectral band calculations on the heat flux distributions.

The schematic of a cylindrical furnace $1m$ in diameter and $3m$ long is shown in Fig. 6.4. A temperature of $T = 1000°C$ is assigned to the furnace interior surface. The furnace volume is filled in with hot gases uniformly mixed at a total pressure of 1 atmosphere and a temperature of $T = 1500°C$. The emissivity of the furnace interior surface is 0.7. To calculate the heat flux distribution at the furnace interior surface we use both constant values, $a = 0.100, 0.125, 0.150, 0.175, 0.200m^{-1}$, for the medium absorption coefficient and the exponential wide model described above. We assume for example that the hot gases filling up the furnace are products of complete combustion of methane with air at 15% excess air level. The composition of combustion products is listed in Tables 7.5. For the potential gas species in this table, the numbers in brackets indicate the absorption band falling completely or partially into the spectral interval.

To carried out the numerical calculations, a subdivision of the spectrum into spectral intervals of fixed width is needed. As explained earlier the width of each spectral interval increases with path length. The band centers (or one band limit) of water and carbon dioxide are known. Consequently, we use the radius (i.e. largest path length) of the furnace to subdivide the entire spectrum into fixed finite spectral intervals. As such, the spectral intervals obtained will include the widest band encountered. The results of this subdivision of the entire spectrum, although arbitrary but useful for engineering calculations, are shown in Table 7.5.

Figure 7.4 shows the predicted incident radiative heat flux at the furnace inner side surface for the range of medium absorption coefficients $a = 0.100, 0.125, 0.150, 0.175, 0.200 \ m^{-1}$. Also shown in this figure are the predictions obtained from spectral calculations. The associated net radiative heat flux at the surface are shown too. We can observe from this figure that the incident heat flux on the furnace interior surface increases almost linearly with the medium absorption coefficient.

Table 7.5. Combustion process of methane with excess air level of 15%. Subdivision of the emission spectrum into spectral intervals.

Fuel and oxidiser

Input	kg	Mass fraction	Molecular weight
CH_4	1.0000	0.0480	16.0430
O_2	4.5875	0.2204	31.9988
N_2	15.2289	0.7316	28.0134

Products of complete combustion

Output	kg	Mass fraction	Mole fraction	Molecular weight
CO_2	2.7432	0.1318	0.0832	44.0100
H_2O	2.2459	0.1079	0.1664	18.0153
O_2	0.5984	0.0287	0.0250	31.9988
N_2	15.2289	0.7316	0.7255	28.0134

Spectral Interval (n)	Spectral limits, μm	Emissive power $W \cdot m^2$	Potential gas species
1	0.0000 − 1.3596	4715.15	none
2	1.3596 − 1.3996	830.42	$H_2O(5)$
3	1.3996 − 1.8262	13068.06	none
4	1.8262 − 1.9200	3654.88	$H_2O(4),CO_2(6)$
5	1.9200 − 2.0308	4530.01	$CO_2(6)$
6	2.0308 − 2.4220	16700.64	none
7	2.4220 − 2.8256	16531.24	$H_2O(3)$
8	2.8256 − 2.9489	4681.29	$H_2O(3),CO_2(5)$
9	2.9489 − 3.1438	6974.79	$CO_2(5)$
10	3.1438 − 4.1494	27722.59	none
11	4.1494 − 4.7202	10600.51	$CO_2(4)$
12	4.7202 − 5.1722	6489.52	none
13	5.1722 − 7.8952	19820.66	$H_2O(2)$
14	7.8952 − 8.8549	3073.06	none
15	8.8549 − 9.7236	1987.56	$CO_2(3)$
16	9.7236 − 10.094	682.18	$CO_2(3,2)$
17	10.094 − 10.644	870.98	$CO_2(2)$
18	10.644 − 11.216	758.60	$H_2O(1),CO_2(2)$
19	11.216 − 12.764	1507.66	$H_2O(1)$
20	12.764 − 18.163	2303.39	$H_2O(1),CO_2(1)$
21	18.163 − $+\infty$	1447.55	$H_2O(1)$

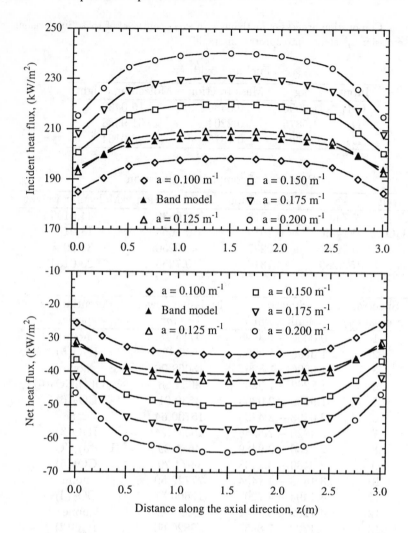

Figure 7.4. Predictions of irradiance and net radiative heat flux distributions at the furnace side wall

For a fixed value of the medium coefficient of absorption, the incident heat flux is minimum near the front and end wall and attains its maximum at the mid-distance along the inner side wall. The absolute value of the net radiative heat flux reaches its maximum at the mid-distance along the side wall. We may observe further that a mean value of $0.125m^{-1}$ for the medium absorption coefficient can be considered as an approximation to the spectral calculations. This is in agreement with the value of $0.11m^{-1}$

Table 7.6. Predictions of irradiance and net radiative heat flux distributions at the furnace side wall

	Incident radiative heat fluxes, kW/m^2					
	constant absorption coefficient, $a\ (m^{-1})$					
$z(m)$	0.100	0.125	0.150	0.175	0.200	$EW\,BM$
0.00	185.06	193.07	200.77	208.16	215.25	194.46
0.25	190.60	200.00	209.04	217.75	226.14	199.66
0.50	195.23	205.65	215.65	225.28	234.54	203.84
0.75	196.84	207.60	217.94	227.87	237.41	205.27
1.00	197.89	208.88	219.42	229.54	239.27	206.27
1.25	198.37	209.45	220.09	230.29	240.09	206.74
1.50	198.52	209.63	220.29	230.53	240.35	206.89
1.75	198.37	209.45	220.09	230.29	240.09	206.74
2.00	197.89	208.88	219.42	229.54	239.27	206.27
2.25	196.84	207.60	217.94	227.87	237.41	205.27
2.50	195.23	205.65	215.65	225.28	234.54	203.84
2.75	190.60	200.00	209.04	217.75	226.14	199.66
3.00	185.06	193.07	200.77	208.16	215.25	194.46

	Net radiative heat fluxes, kW/m^2					
	constant absorption coefficient, $a\ (m^{-1})$					
$z\,(m)$	0.100	0.125	0.150	0.175	0.200	$EW\,BM$
0.00	−25.27	−30.88	−36.27	−41.44	−46.41	−31.86
0.25	−29.15	−35.74	−42.06	−48.16	−54.03	−35.49
0.50	−32.40	−39.69	−46.69	−53.43	−59.91	−38.42
0.75	−33.52	−41.05	−48.29	−55.24	−61.92	−39.42
1.00	−34.26	−41.95	−49.33	−56.42	−63.22	−40.13
1.25	−34.59	−42.35	−49.79	−56.94	−63.80	−40.45
1.50	−34.70	−42.48	−49.94	−57.10	−63.98	−40.56
1.75	−34.59	−42.35	−49.79	−56.94	−63.80	−40.45
2.00	−34.26	−41.95	−49.33	−56.42	−63.22	−40.13
2.25	−33.52	−41.05	−48.29	−55.24	−61.92	−39.42
2.50	−32.40	−39.69	−46.69	−53.43	−59.91	−38.42
2.75	−29.15	−35.74	−42.06	−48.16	−54.03	−35.49
3.00	−25.27	−30.88	−36.27	−41.44	−46.41	−31.86

obtained using the mean beam length concept. The numerical values of incident and net heat flux accompanying the foregoing figures are listed in Tables 7.6.

The numerical simulations were performed using 254 nodal points, hence 504 triangular surface elements, placed over the furnace interior surface. We use, per triangular surface elements, 75 quadratures points for inner integrals and 4 quadrature points for outer integrals. For total spectral calculations using the exponential wide band model, the results (over the whole spectrum) presented hereafter took 81 minutes and 48 seconds CPU time of a machine equipped with a Pentium II processor 400 MHz.

8

Application to Industrial Furnace

In this chapter we go on to consider an industrial furnace situation. A furnace consists of a heat source (burner producing a flame), a heat sink (the tube bank of a boiler or heat exchanger, or cooling tubes), and the refractory walls. The discussion here is on application of the developed method to the classical radiative heat transfer problem in thermal engineering:

Given a furnace in which the temperature distribution is known, it is required to characterize the radiant heat exchange, specifically to calculate the incident heat flux to the load (heat sink).

The semi-industrial scale furnace considered in this chapter is shown in Fig. 8.1. The experimental data reported here were extracted from the M2-Trials performed at the International Flame Research Foundation facilities in IJmuiden [135]. The M2-Trials were conducted at IJmuiden in the IFRF Furnace No.1 to investigate the influence of operational variables - fuel type, burner type, excess air, oxygen enrichment of combustion air, and degree of swirl in the combustion air - on heat transfer in the furnace. The data were obtained in a sufficient detail to assess available mathematical models for the prediction of temperature and radiative heat flux distributions in a semi-industrial size burner/furnace system. The trials aimed at providing information at a practical level for immediate use in furnace engineering and at a theoretical level for further optimization and validation of mathematical models.

Figure 8.1. Semi-industrial scale furnace at IJmuiden

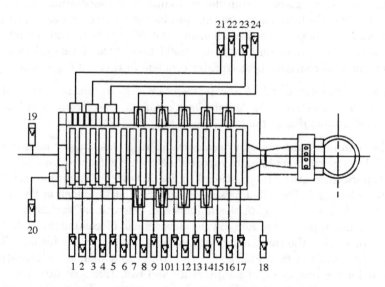

1-17 : Double loop cooling pipe   20   : Support rail of hot and cold target
18   : Vertical measuring slot (east wall) 21-23 : Horizontal slot's (west wall)
19   : Burner   24   : Vertical slot's (west wall)

Figure 8.2. Distribution of cooling loops in the furnace as used during the M2-Trials

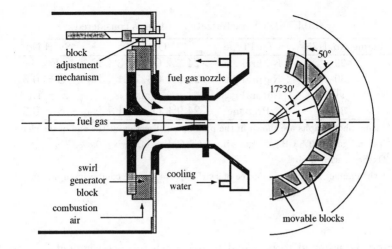

Figure 8.3. Burner equiped with a movable block type swirl generator

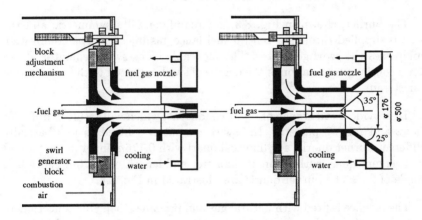

Figure 8.4. Design of the cylindrical and divergent burner heads

8.1 Experimental Configuration

In this last exercise the experimental furnace considered is a horizontal tunnel type furnace, Fig. 8.1-8.2, of $2m \times 2m$ cross-section and a length of $6.25m$. The furnace consists of a front wall, seventeen segments approximately square in cross section with an arch roof, and a furnace end wall with a chimney opening. The burner, firing along the furnace axis, is mounted on the furnace front wall. The front wall, the segments and the furnace end wall are constructed out of refractory bricks. The emissivity of the re-

Table 8.1. Characterization of flames considered

Flame No. [135][1]	Fuel type	Oxidant[3]	Swirl No.[3]
29	Natural gas[2]	24.7% O_2, 75.3% N_2	$S = 0$
30	Natural gas[2]	24.9% O_2, 75.1% N_2	$S = 0.5$
11	Natural gas[2]	air	$S = 1.95$
31	Propane	24.9% O_2, 75.1% N_2	$S = 0.5$

[1]Flame photographs are shown in the photo-pannel section at the end of the chapter
[2]Natural gas: 81.3% CH_4, 2.9% C_2H_6, 0.6% C_3H_8, 14.4% N_2, 0.8% CO_2
[3]Oxygen enriched air
[4]Swirl number indicating a degree of swirl imparted to the combusin air stream

fractory is assumed to be equal to 0.5. Each furnace segment is partially covered by two cooling loops separated by a distance of $0.185m$. The cooling loops, with a surface emissivity of 0.8, are located equidistantly from the edge and center of each $0.370m$ wide segment.

The burner, shown in Fig. 8.3, is circumferentially symmetric and the fuel is supplied through a central fuel lance gas injector. The combustion air passes the swirl generator. The moveable block swirl generator allows for a continuous variation of the degree of swirl imparted to the combustion air stream.

To provide a wide range of input conditions, which will result in different flows and mixing patterns, the swirl generator is fitted to two basically different burner quarls: a cylindrical quarl with $0.176m$ inner diameter and a divergent quarl with a throat diameter and length of $0.176m$ and a half angle of $25°$. The burner quarls are illustrated in Fig. 8.4.

The furnace is fired with natural gas and propane [135]. Our concern here is to characterize the radiative heat exchanges in the furnace and to investigate the influence of fuel and flame types on the heat flux distributions. To this end, three natural gas flames and a propane flame are examined. Their composition is given in Table 8.1.

The flames are of 3 MW thermal input and the excess air level is 6%. This thermal input is large enough to provide a reasonable representation of industrial scale flames, whilst being small enough for research purposes. Measurements of gas temperature (degrees Celsius) are carried out inside the furnace at various traverses. They are listed in Table 8.2 through to and including Table 8.5 at the end of this chapter. The mean values of the hourly recorded furnace interior surface temperatures for the five flames produced are plotted in Fig. 8.5 and are listed in Table 8.6.

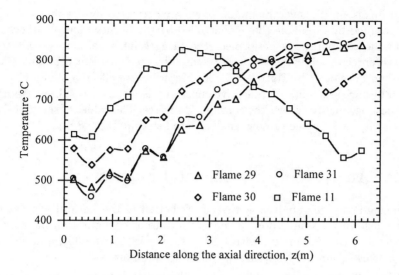

Figure 8.5. Temperature measurements along the furnace inner side wall

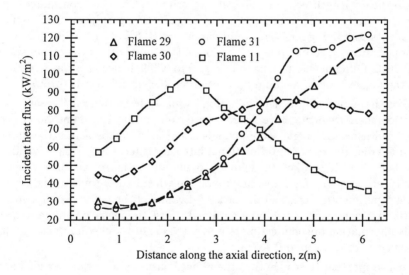

Figure 8.6. Incident heat flux measurements at the furnace inner side wall

To characterize the radiative heat exchanges, measurements of incident radiative heat flux at the interior surface, along the furnace axis, are carried out by using a hemispherical radiometer. Their values are plotted in Fig. 8.6 and are listed in Table 8.7. The flames possess very different temperature distributions despite the same thermal input. Subsequently, the incident heat fluxes and the overall amount of energy extracted by the heat sink are determined by the flame shape controlled by the burner fluid dynamics.

In the computations performed, the furnace inner refractory wall and the seventeen cooling loops are represented by an equivalent heat sink. The method developed by Hottel [78, 136] prescribes an effective emissivity to each furnace segment. The effective emissivity is merely a weighted average of the true emissivity over the considered surface. The values of the effective emissivity used in the present analysis are listed in Table 8.8.

8.2   Analysis of Experimental Flames

We commence the analysis with the prediction of the natural gas flames obtained with a simple double concentric burner configuration in the absence of swirl (Flame 29), with a moderate swirl (Flame 30), and with a high swirl (Flame 11), applied to the combustion air.

To carry out the computations of radiative exchanges, there is a need for determining values of the medium absorption coefficient. The mean values of the medium absorption coefficient can be estimated using Hottel's mean beam length concept. For the experimental furnace considered the mean beam length is equal to $1.55m$. Using Hottel's graphs for combustion products containing 20.4% of water vapor and 10.5% of carbon dioxide at a temperature of 1200°C, one may estimate the emissivity of the combustion products filling in the furnace to be equal to 0.209, hence an approximation of the mean value of medium absorption coefficient follows as $0.15m^{-1}$.

For the experimental determination of local absorption coefficients, radiation measurements have to be carried out inside the flame body. To this end a narrow angle radiometer is traversed in steps across the flame and the distribution of uni-directional intensity is recorded. This traversing method [137, 138] allows a point by point estimation of the mean emissivity/absorptivity for finite flame volume within the furnace. The available experimental data contain measured values of the traverse emissivities [135]. A mean value of 0.24 was obtained across the furnace volume. Using this information one can estimate a mean value of the medium absorption coefficient to be around $0.137m^{-1}$.

It is important to realize that the above considerations can provide only a rough estimation of the mean absorption coefficient. The absorption coefficient actually varies with the position in the flame and can be strongly altered by the presence of soot.

8.2.1   The Non-Swirling Natural Gas Flame

The non-swirling natural gas flame, Flame 29, is produced using the cylindrical burner of Fig. 8.4 and a single hole gas injector. The thermal input

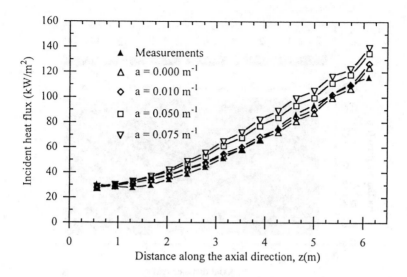

Figure 8.7. Flame 29; comparison of measured and predicted incident heat fluxes at the furnace side wall

into the furnace is 3 MW. The heat sink, represented by the cooling loops, extracts 1.23 MW energy while the enthalpy of combustion products, at temperature 1184°C, at the furnace exit is 1.52 MW. Thus, around 0.25 MW of energy is dissipated through the furnace refractory walls.

Visually the flame is almost $5m$ long and resembles a free jet, see photo panels section. For this flame the distribution of gas temperature throughout the furnace volume is listed in Table 8.2.

The gas temperature rises continuously from the furnace front wall towards the end wall. The same observation is made for the measured incident heat flux of radiation at the side wall.

Figure 8.7 shows the predictions of incident radiative heat fluxes for the values of absorption coefficients: $a = 0.0m^{-1}$ (transparent medium), $a = 0.01m^{-1}$, $a = 0.05m^{-1}$ and $a = 0.075m^{-1}$. We can observe that the agreement between measured and computed values of incident fluxes is obtained with small values of medium absorption coefficients (e.g. $a \leq 0.010m^{-1}$). Thus the gaseous medium in this case is strongly transparent. The net radiative heat flux distribution $q_V(p)$ inside the furnace volume, in kW/m^3, is plotted in Fig. 8.8. Here, the calculations have been performed with a medium absorption coefficient equals to $0.010m^{-1}$. As it was the case for the incident radiative heat flux, we observe from Fig. 8.8 that the net heat flux inside the medium is minimum (in absolute value) near the front wall reflecting the lower gas temperature, interior surface temperature and irradiance at the furnace wall in this region. It increases with the distance along the furnace firing axis as a result of the increase in the gas and interior surface temperature, and irradiance $H(p)$ on the furnace interior surface.

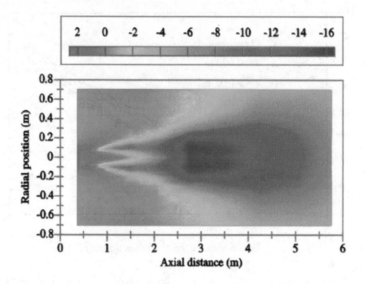

Figure 8.8. Flame 29; predicted net heat source distribution inside the furnace

We may also observe that the flame volume is small compare to the entire furnace volume.

Having determined the absorbed heat flux distribution $q_V^a(p)$, the total radiative energy absorbed by the gases filling in the furnace volume Q_V^a and the total radiative energy emitted by the gases filling in the furnace volume Q_V^e can be calculated. It is noteworthy that there are no physical device that can be used to measure these quantities inside an enclosure. Thus one must rely on numerical predictions and the numerical procedure used is the only element of confidence.

For the non-swirling natural gas flame, the energy absorbed by the volume of gas amounts to 0.09 MW, for $a = 0.010m^{-1}$. The energy emitted by the volume of gas is in the range of 0.158 MW, for the same value $a = 0.010m^{-1}$ of medium absorption coefficient. The above analysis reveals that energy is transferred to the heat sink mainly by convection and by radiation from the furnace walls.

8.2.2   The Low-Swirling Natural Gas Flame

Flame 30 differs from Flame 29 only by the swirl imparted to the combustion air. This has an important effect: it reduces the flame length by producing higher rates of mixing closer to the nozzle exit.

As shown in the photo panel section, applying a weak swirl ($S = 0.5$) to the combustion air changes the flow and mixing pattern. This causes a broadening of the flame and reduces the flame length to approximately

3.5m. For the thermal input of 3 MW, the heat sink absorbs 1.43 MW energy while the enthalpy of combustion products at the furnace exit amounts to 1.3 MW. Around 0.27 MW of the energy is dissipated through the furnace refractory walls. The furnace exit gas temperature is 1029°C, a value that is substantially lower than that of the non-swirling flame. The application of swirl results in a 16% increase in the amount of energy transferred to the heat sink.

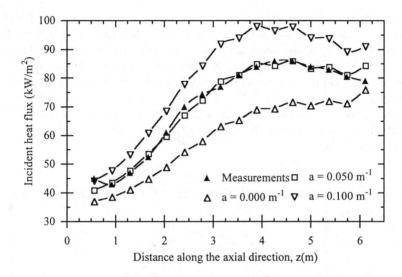

Figure 8.9. Flame 30; comparison of measured and predicted incident heat fluxes at the furnace side wall

Comparisons between the measured and predicted incident heat flux along the furnace wall are illustrated in Fig. 8.9, for the range of absorption coefficients $a = 0.000, 0.050, 0.100m^{-1}$. Applying a swirl to the combustion air results in a radial spread of the flame. The volume of hot radiating gases increases, and we have a substantial increase of the incident heat flux at the furnace wall. In this case, a good agreement between the measured and predicted incident heat flux on the furnace interior surface is obtained with the medium coefficient of absorption $a = 0.050m^{-1}$.

Our computation predict that the total energy absorbed by the volume of gases (for the coefficient of absorption $a = 0.050m^{-1}$) is $\hat{Q}_V^a = 0.560$ MW, which corresponds to 18.7% of the thermal input. The energy emitted by the volume of gas amounts to 1.281 MW which corresponds to 42.7% of the thermal input.

8.2.3   The High-Swirling Natural Gas Flame

The high-swirling natural gas flame ($S = 1.95$), Flame 11, is produced
upon introducing the fuel gas at the burner throat using a 16 divergent hole
gas injector. The partly penetrating gas jets interact with a swirl induced
reverse flow zone which is formed inside the furnace and just downstream
of the burner quarl, producing a very intense $0.5m$ long flame, see photo
panels section.

The distribution of gas temperature inside the furnace volume for this
flame is shown in Table 8.4. We can observe that the flame volume has
increased significantly if compared with the flame volume of both the non-
swirling (Flame 29) and moderately swirling (Flame 30) natural gas flames.
For the thermal input considered (3 MW), the energy absorbed by the heat
sink has increased to 1.50 MW while the flue gas enthalpy has decreased
to 1.08 MW. The furnace exit gas temperature has dropped to 779°C. The
heat dissipation through the furnace refractory walls has increased to a
value of 0.45 MW.

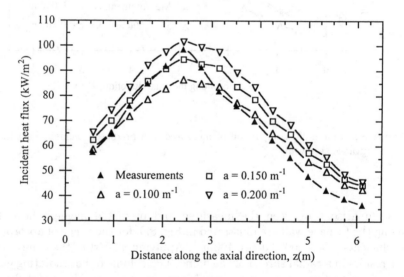

Figure 8.10. Flame 11; comparison of measured and predicted incident heat fluxes
at the furnace side wall

A comparison between the measured and computed incident radiative
heat fluxes is shown in Fig. 8.10. The incident heat flux curves clearly
indicate the presence of a short flame with the highest flame temperature
located at approximately $2m$ distance downstream of the furnace front wall.

8.2.4 The Non-Swirling Propane Flame

Using the same double concentric burner configuration as for the non-swirling natural gas flame, Flame 29, but with a single hole gas injector, a non-swirling propane flame, Flames 31, is produced. Flame 29 and 31 have a very similar flow pattern, see photo panel section. The heat transfer characteristics for both flames are similar. The thermal input into the furnace remains 3 MW and the heat sink extracts 1.25 MW energy. The enthalpy of combustion products at temperature 1200°C at the furnace exit is 1.43 MW and around 0.32 MW of energy is dissipated through the furnace refractory walls.

The products of complete combustion of propane with the enriched air contain 13% of carbon dioxide and 17.4% of water vapor. For the furnace exit temperature of 1200°C, one may estimate the emissivity of the combustion products filling in the furnace volume to be equal to 0.23, which corresponds to a value of $0.168 m^{-1}$ for the mean absorption coefficient. However, the traversing method measurements [135] indicate that the coefficient should take a value of $0.11 m^{-1}$ in the first three meters of the furnace length and $0.2 m^{-1}$ downstream.

Figure 8.11 shows the comparison between the measured and predicted incident radiative heat fluxes at the furnace wall. The overall analysis of the propane flame indicates that its radiative characteristics are similar to those of the non-swirling natural gas flame (Flame 29).

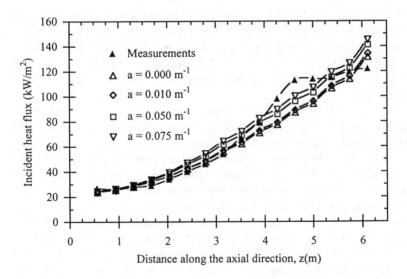

Figure 8.11. Flame 31; comparison of measured and predicted incident heat fluxes at the furnace side wall

8.3   Concluding Remarks

The foregoing discussion was meant to elucidate the ability of our algorithm to accurately predict the radiative heat transfer in flames and furnaces. To investigate the influence of the aerodynamic flow patterns and fuel type on the heat flux distribution within the furnace, four flames have been studied. Changes in the burner configuration and/or fuel type alter the combustion pattern and the objective was to show how these changes influence the radiative exchange. It has been shown that the swirl imparted into the combustion air stream has a major impact on the incident heat flux distribution to the furnace heat sink. With increasing the swirl, the radial spread of the flames increases. This results in an increase of both the flame volume and incident radiative heat flux to the heat sink.

The computations in this chapter were performed with 218 points placed over the furnace wall for each case. This resulted in 432 triangular surface elements in partitioning the furnace wall into finite elements. We have employed per surface elements, 27 quadratures points for inner integrals and 4 quadrature points for outer integrals. Further, we have used the trapezoidal rule with equally space points and we have applied $2cm$ spacing for calculation of the line integrals throughout the furnace volume. The results presented here took approximately 6 minutes per run of a PC equipped with a Pentium II processor 400 MHz; 2 minutes for the calculation of the matrix B of the linear system $B \cdot H_{\epsilon,h(n)} = L$, approximately 2.4 minutes for the calculation of the vector L, and 1.6 minutes for solving the linear system $B \cdot H_{\epsilon,h(n)} = L$ of size 218×218 using Cholesky method.

The algorithm developed has been used together with the experimental data to examine similarity between natural gas and propane flames. While analyzing the influence of flow pattern on the heat flux distributions, it has been established that the application of a swirl to the combustion air has a much higher influence on the radiative fluxes than the change from natural gas to propane.

In the examples studied in this chapter we have used several values for the medium absorption coefficient. These values were not coincidental. As indicated already, for products of complete combustion of methane (6% excess air level) at 1200°C temperature, the total emissivity can be estimated to be around 0.209. This corresponds to the mean absorption coefficient of $0.15m^{-1}$. Generally, the higher the temperature the lower the values of absorption coefficient. Therefore, it is expected that values up to $0.1m^{-1}$ are applicable for non-sooty gas flames. Our considerations have clearly indicated that for reliable prediction of radiative exchanges, an appropriate value of the absorption coefficient should be calculated locally. A proper determination of the absorption coefficient which varies locally with the temperature and medium composition, however, is a complex task which requires a comprehensive investigation.

8.4 Photo Panels - Experimental Flames

Figure 8.12. Flame 29; non-swirling natural gas flame (S=0)

Figure 8.13. Flame 30; low-swirling natural gas flame (S=0.5)

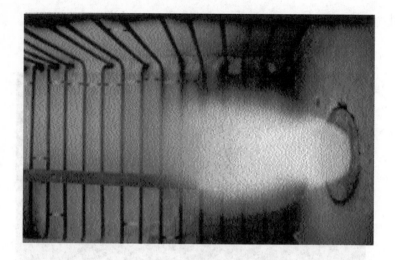

Figure 8.14. Flame 11; high swirling natural gas flame (S=1.95)

Figure 8.15. Flame 31; non-swirling propane flame (S=0)

8.5   In-Flame and Wall Measurements

Table 8.2. Flame 29; temperature measurements ($^\circ C$) inside the furnace

$r(m)$	axial direction $z(m)$							
	0.370	0.555	0.740	0.925	1.110	1.295	1.480	-
0.000	75	100	205	355	480	615	730	—
0.025	89	110	220	390	540	660	795	—
0.050	85	190	405	605	710	830	915	—
0.075	125	180	560	915	990	1045	1025	—
0.100	260	255	385	855	1100	1170	1200	—
0.125	425	345	350	590	910	1065	1160	—
0.150	570	450	390	465	690	850	1030	—
0.175	610	525	475	467	606	730	875	—
0.200	630	595	565	470	523	610	730	—
0.300	630	610	605	590	570	560	580	—
0.400	630	620	605	600	595	600	595	—
0.500	630	615	600	595	595	595	595	—
0.600	630	610	595	595	585	585	575	—
0.700	630	605	595	595	575	575	565	—
1.000	495	485	503	521	516	510	543	—

$r(m)$	axial direction $z(m)$							
	1.665	1.850	2.035	2.750	3.500	4.250	5.000	5.750
0.000	850	972	1065	1370	1370	1325	1285	1215
0.025	885	985	1088	1372	1370	1327	1285	1215
0.050	980	1050	1135	1375	1380	1330	1285	1215
0.075	1090	1130	1182	1383	1380	1327	1285	1215
0.100	1150	1170	1195	1385	1380	1325	1285	1215
0.125	1120	1150	1175	1377	1370	1318	1280	1212
0.150	1017	1085	1117	1370	1360	1317	1275	1210
0.175	880	975	1090	1350	1350	1317	1272	1207
0.200	770	880	950	1330	1340	1310	1270	1205
0.300	620	670	702	1075	1220	1270	1245	1190
0.400	610	632	660	880	1080	1210	1200	1170
0.500	610	640	655	810	985	1150	1160	1150
0.600	610	635	670	790	955	1095	1125	1130
0.700	585	595	655	790	925	1050	1080	1110
1.000	575	568	560	640	705	775	815	835

Table 8.3. Flame 30; temperature measurements ($°C$) inside the furnace

$r(m)$	axial direction $z(m)$							
	0.370	0.555	0.740	0.925	1.110	1.295	1.480	-
0.000	115	280	990	910	988	1115	1220	—
0.025	175	610	880	920	1010	1135	1220	—
0.050	325	620	700	925	1045	1155	1275	—
0.075	240	650	635	960	1075	1167	1235	—
0.100	300	530	615	965	1090	1180	1240	—
0.125	400	465	607	938	1065	1132	1235	—
0.150	510	460	600	910	1040	1132	1230	—
0.175	610	500	612	842	990	1131	1200	—
0.200	600	540	625	785	940	1085	1170	—
0.300	760	718	675	735	790	910	1000	—
0.400	755	788	738	730	760	825	980	—
0.500	750	780	765	750	789	815	870	—
0.600	745	824	765	760	890	825	865	—
0.700	740	868	765	760	790	810	855	—
1.000	561	540	559	577	579	580	616	—

$r(m)$	axial direction $z(m)$							
	1.665	1.850	2.035	2.750	3.500	4.250	5.000	5.750
0.000	1280	1320	1340	1360	1320	1320	1190	1120
0.025	1285	1320	1342	1362	1320	1320	1187	1120
0.050	1290	1320	1345	1365	1320	1320	1185	1120
0.075	1290	1320	1340	1362	1320	1320	1180	1120
0.100	1290	1315	1335	1360	1320	1320	1175	1120
0.125	1275	1307	1325	1351	1315	1298	1171	1117
0.150	1260	1300	1315	1342	1310	1277	1167	1115
0.175	1245	1285	1305	1334	1005	1256	1163	1112
0.200	1230	1270	1295	1325	1300	1235	1160	1110
0.300	1085	1170	1180	1290	1270	1220	1150	1100
0.400	970	1040	1075	1230	1225	1207	1140	1195
0.500	910	970	1000	1150	1190	1195	1120	1075
0.600	885	930	960	1085	1125	1092	1100	1060
0.700	875	910	935	1060	1115	1090	1075	1055
1.000	652	656	660	750	790	800	802	745

Table 8.4. Flame 11; temperature measurements (°C) inside the furnace

$r(m)$	\multicolumn{8}{c}{axial direction $z(m)$}
	0.370	0.555	0.740	0.925	1.110	1.295	1.480	-
0.000	1590	1490	1480	1430	1385	1340	1290	−
0.025	1550	1495	1472	1427	1380	1336	1286	−
0.050	1510	1500	1465	1425	1377	1332	1282	−
0.075	1470	1490	1457	1417	1373	1329	1278	−
0.100	1440	1480	1450	1410	1370	1325	1275	−
0.125	1310	1430	1435	1390	1351	1311	1271	−
0.150	1158	1386	1425	1370	1332	1297	1267	−
0.175	986	1310	1390	1350	1313	1283	1253	−
0.200	815	1235	1325	1330	1295	1270	1240	−
0.300	820	990	1125	1175	1105	1165	1170	−
0.400	790	845	960	1035	1060	1070	1080	−
0.500	780	680	810	910	950	985	1010	−
0.600	770	722	790	850	890	930	970	−
0.700	760	765	789	810	860	920	960	−
1.000	613	610	646	681	696	710	745	−

$r(m)$	\multicolumn{8}{c}{axial direction $z(m)$}
	1.665	1.850	2.035	2.750	3.500	4.250	5.000	5.750
0.000	1260	1215	1190	1110	1000	915	860	790
0.025	1257	1213	1190	1110	1000	913	856	792
0.050	1255	1212	1190	1110	1000	912	853	793
0.075	1252	1211	1190	1110	1000	911	850	794
0.100	1250	1210	1190	1110	1000	910	847	795
0.125	1243	1202	1186	1108	1000	910	847	795
0.150	1237	1195	1182	1107	1000	911	846	796
0.175	1231	1187	1178	1106	1000	911	846	796
0.200	1225	1180	1175	1105	1000	912	846	797
0.300	1160	1140	1145	1100	1000	915	845	800
0.400	1115	1100	1110	1097	1010	922	862	795
0.500	1040	1065	1080	1095	1020	930	880	790
0.600	1000	1040	1060	1080	1020	925	870	785
0.700	995	1030	1055	1065	1020	920	860	780
1.000	780	780	780	820	775	718	645	561

Table 8.5. Flame 31; temperature measurements ($^\circ C$) inside the furnace

$r(m)$	0.370	0.555	0.740	0.925	1.110	1.295	1.480	-
				axial direction $z(m)$				
0.000	270	310	450	550	660	765	855	—
0.025	310	350	460	575	670	775	870	—
0.050	330	320	763	810	760	865	975	—
0.075	135	370	1050	1190	1130	1100	1120	—
0.100	230	230	400	710	1040	1200	1235	—
0.125	395	310	330	400	640	960	1130	—
0.150	570	470	360	400	510	715	915	—
0.175	620	540	429	425	500	560	715	—
0.200	640	690	490	450	490	510	600	—
0.300	630	650	600	560	520	520	620	—
0.400	630	610	600	590	560	530	582	—
0.500	625	610	590	580	560	540	545	—
0.600	620	610	585	570	560	540	545	—
0.700	615	600	580	560	560	540	560	—
1.000	483	460	486	512	506	500	540	—

$r(m)$	1.665	1.850	2.035	2.750	3.500	4.250	5.000	5.750
				axial direction $z(m)$				
0.000	845	1005	1070	1240	1325	1250	1220	1175
0.025	980	1020	1080	1245	1325	1250	1217	1183
0.050	1025	1075	1130	1250	1325	1255	1215	1190
0.075	1140	1160	1200	1263	1325	1250	1215	1183
0.100	1250	1245	1200	1275	1325	1245	1215	1175
0.125	1249	1265	1240	1283	1323	1240	1212	1174
0.150	1090	1180	1250	1290	1320	1235	1208	1173
0.175	910	1050	1150	1290	1312	1230	1204	1172
0.200	750	895	1025	1280	1305	1225	1200	1170
0.300	550	605	650	1060	1210	1170	1170	1155
0.400	555	603	640	840	1070	1120	1135	1140
0.500	560	600	630	760	960	1040	1110	1125
0.600	550	600	630	740	890	1000	1075	1095
0.700	540	600	630	735	860	960	1035	1090
1.000	580	570	560	660	750	805	840	845

Table 8.6. Measured inner wall temperature ($°C$)

Location $P(r, z)$		Flame number			
$r(m)$	$z(m)$	29	30	31	11
1.000	0.000	504	581	506	616
1.000	0.185	504	581	506	616
1.000	0.555	485	540	460	610
1.000	0.925	521	577	512	681
1.000	1.295	510	580	500	710
1.000	1.665	575	652	580	780
1.000	2.035	560	660	560	780
1.000	2.405	628	724	652	827
1.000	2.775	640	750	660	820
1.000	3.145	693	784	728	812
1.000	3.515	705	790	750	775
1.000	3.885	749	806	788	736
1.000	4.255	775	800	805	718
1.000	4.625	804	816	836	683
1.000	4.995	815	802	840	645
1.000	5.365	826	724	850	615
1.000	5.735	835	745	845	561
1.000	6.105	841	775	865	578
1.000	6.290	1098	838	1117	578
Flue gas		1184	1029	1201	779

Table 8.7. Measured incident heat flux at the inner wall (kW/m^2)

Location $P(r, z)$		Flame number			
$r(m)$	$z(m)$	29	30	31	11
1.000	0.555	30.5	45.0	27.0	57.5
1.000	0.925	28.5	43.0	26.0	65.0
1.000	1.295	28.0	47.0	27.5	76.0
1.000	1.665	30.0	52.5	29.0	85.0
1.000	2.035	34.5	61.0	34.0	92.0
1.000	2.405	39.0	70.0	40.0	98.5
1.000	2.775	44.5	74.5	46.0	91.5
1.000	3.145	51.5	77.0	54.0	82.0
1.000	3.515	58.5	81.0	67.5	76.0
1.000	3.885	66.0	84.0	80.0	70.0
1.000	4.255	76.0	86.0	98.0	62.5
1.000	4.625	86.5	86.0	113.0	55.5
1.000	4.995	94.0	84.0	114.0	48.0
1.000	5.365	102.5	83.0	115.0	42.5
1.000	5.735	110.5	80.5	120.0	39.0
1.000	6.105	116.0	79.0	122.0	36.5

Table 8.8. Effective emissivity distribution at the furnace wall

$P(r,z)$		refractory		cooling pipe		effective
$r(m)$	$z(m)$	area	emissivity	area	emissivity	emissivity
0.000	0.000	–	–	–	–	0.637
0.185	0.000	–	–	–	–	0.500
0.370	0.000	–	–	–	–	0.500
0.555	0.000	1.290	0.500	–	–	0.514
0.740	0.000	1.290	0.500	–	–	0.514
1.000	0.000	2.051	0.500	–	–	0.514
1.000	0.185	2.396	0.500	1.653	0.800	0.659
1.000	0.555	2.396	0.500	1.653	0.800	0.659
1.000	0.925	2.396	0.500	1.653	0.800	0.659
1.000	1.295	2.396	0.500	1.653	0.800	0.659
1.000	1.665	2.396	0.500	1.653	0.800	0.659
1.000	1.665	2.396	0.500	1.653	0.800	0.659
1.000	2.035	2.396	0.500	1.653	0.800	0.659
1.000	2.405	2.580	0.500	1.571	0.800	0.678
1.000	2.775	2.368	0.500	1.571	0.800	0.650
1.000	3.145	2.580	0.500	1.571	0.800	0.678
1.000	3.515	2.368	0.500	1.571	0.800	0.650
1.000	3.885	2.580	0.500	1.571	0.800	0.678
1.000	4.255	2.368	0.500	1.571	0.800	0.650
1.000	4.625	2.580	0.500	1.571	0.800	0.678
1.000	4.995	2.368	0.500	1.571	0.800	0.650
1.000	5.365	2.580	0.500	1.571	0.800	0.678
1.000	5.735	2.368	0.500	1.571	0.800	0.650
1.000	6.105	2.580	0.500	1.571	0.800	0.678
1.000	6.290	2.368	0.500	–	–	0.500
0.740	6.290	2.150	0.500	–	–	0.500
0.555	6.290	2.150	0.500	–	–	0.500
0.370	6.290	–	–	–	–	0.900
0.185	6.290	–	–	–	–	0.900
0.000	6.290	–	–	–	–	0.900

9

Radiation in Scattering Media

The preceding chapters have dealt with radiative heat transfer in non-scattering media only. This assumption is justifiable for many engineering radiative heat transfer calculations where the system under consideration is not to depart arbitrarily far from the non-scattering assumption. Now the question is: what happens when an enclosure contains particles suspended in the medium and scattering becomes of considerable importance? In this chapter we consider systems where the effects of scattering mechanism onto the overall radiative heat exchange is not negligible.

9.1 Formulation of the Problem

In absorbing, emitting, and scattering media, the intensity of incident radiation is described by (2.2), which we recall:

$$I^i(r,p) \;=\; I^o(r)\tau(r,p) + \int_{(r,p)} \beta(u)G(r,u)\tau(u,p)dL(u) \, ,$$

$$G(r,u) \;=\; \left(1 - \frac{\sigma(u)}{\beta(u)}\right) I_b(u) + \frac{\Sigma(r,u)}{\beta(u)} \, ,$$

$$\Sigma(r,u) \;=\; \frac{\sigma(u)}{4\pi} \int_{4\pi} I^i(r,u)\Phi(s,r,u)d\omega(udS(s)) \, .$$

This equation shows that the intensity $I^i(r,p)$ of incident radiation at a moving point p is made up of two terms: the first is the intensity of outgoing radiation at the origin r and into the direction $\overrightarrow{rp}$, attenuated according to Lambert's law along the path (r, p); the second term accounts for the contributions from each intervening radiating element within the enclosure volume. This later term includes spontaneous emission and incoming scattering along the path, reduced by exponential attenuation.

The main difficulty in the relation above stems from the presence of the phase function for scattering, the integrand in the integral term $\Sigma(r, u)$. The process of scattering is represented here by replacing the intensity of blackbody radiation $I_b(u)$ by the source intensity $G(r, u)$ and the absorption coefficient $a(u)$ by the extinction coefficient $\beta(u)$. In scattering media, we may formulate the radiant heat exchange problem as the interrogation:

how much radiation coming from $\overrightarrow{su}$ direction is scattered by a particle at the point u, into $\overrightarrow{ru}$ direction?

The answer to this question is given in terms of a dimensionless phase function for scattering, $\Phi(s, r, u)$. The phase function $\Phi(s, r, u)$ may depend upon the angle of incidence as well as the angle of reflection of radiation.

The idea now is to take advantage of the theoretical developments given in the foregoing chapters. To this end we shall consider the average value $\bar{\Phi}(r, u)$ of the population of phase functions $\Phi(s, r, u)$ over all incident solid angles $\omega(u, dS(s))$ subtended by the surface $dS(s)$ at the point u and defined to be:

$$\bar{\Phi}(r, u) = \frac{1}{4\pi} \int_{4\pi} \Phi(s, r, u) d\omega(u, dS(s)) .$$

The quantity $\bar{\Phi}(r, u)$ is a measure of how much a particle at the location u within the medium scatters into $\overrightarrow{ru}$ direction radiation propagating from all other directions. Then, we replace the phase function $\Phi(s, r, u)$ into the integral term of the source function $G(r, u)$ by the $\bar{\Phi}(r, u)$. Accordingly, the source function may be written in the form:

$$G(r, u) = \left(1 - \frac{\sigma(u)}{\beta(u)}\right) I_b(u) + \frac{\Phi_\sigma(r, u)}{4\pi\beta(u)} \int_{4\pi} I^i(s, u) d\omega(u, dS(s)) ,$$

with $\beta(u) = a(u) + \sigma(u)$ being the medium extinction coefficient, and $\Phi_\sigma(r, u) = \sigma(u)\bar{\Phi}(r, u)$.

The radiative heat transfer problem now is merely displaced. Still, it remains in the realm of radiant energy exchanges owing to the uncertainties associated with the determination of the phase function for scattering $\Phi(s, r, u)$.

For isotropic scattering, $\Phi(s, r, u)$ is equal to unity, hence

$$\bar{\Phi}(r, u) = 1.$$

If scattering is anisotropic, the function $\bar{\Phi}(r, u)$ is not necessarily identically equal to unity.

Let us define the contribution of the participating medium to be:

$$L(r,p) = \int_{(r,p)} \beta(u)G(r,u)\tau(u,p)dL(u) .$$

Then, in an enclosure having an opaque and Lambert's interior surface, the radiative heat balance relations on a unit surface read:

For $\varepsilon(p) = 0$ almost everywhere,

$$q = 0.$$

For $\varepsilon(p) = 1$ almost everywhere,

$$q(p) = e_b(p) - \int_S \left(e_b(r)\tau(r,p) + \pi L(r,p)\right) K(r,p)dS(r) .$$

For $0 < \varepsilon(p) < 1$,

$$\frac{q(p)}{\varepsilon(p)} + \int_S \left(\left(e_b(r) - \rho(r)\frac{q(r)}{\varepsilon(r)}\right)\tau(r,p) + \pi L(r,p)\right) K(r,p)dS(r) = e_b(p),$$

with $\rho(r) = 1 - \varepsilon(r)$ being the reflectivity at the point r.

The heat source distribution q_V inside the enclosure volume vanishes for purely scattering medium. For enclosures filled in with a scattering medium having non zero absorption coefficient, it satisfies the relation:

$$\frac{q_V(p)}{a(p)} = 4e_b(p) - \int_S \left(J(r)\tau(r,p) + \pi L(r,p)\right) K_V(r,p)dS(r) .$$

It is noteworthy that a good description of heat transfer by radiation in absorbing-emitting and scattering medium requires that the extinction coefficient be non-uniform but varies with temperature and/or medium composition as radiation travels throughout the medium. Investigation of the net heat balance relations described hereabove, under the hypothesis of known temperature distributions, absorption coefficient $a(u) \neq 0$ and randomly oriented scattering particles inside the medium, shows that we need concern ourselves with the following problem:

Given an enclosure of opaque interior surface S having directional property effects sufficiently unimportant that it can be treated as a Lambert surface, we seek a function $H_\varepsilon(p)$ continuous over S and such that:

$$H_\varepsilon(p) - \int_S N(r,p)H_\varepsilon(r)dS(r) = \sqrt{1-\varepsilon(p)}\,(F_1(p) - F_2(p)),$$

where $N(r,p)$ is the symmetric kernel already defined,

$$F_1(p) = \frac{1}{4}\int_S\int_{(r,p)} \Phi_\sigma(r,u)\tau(u,p)\frac{q_V(u)}{a(u)}K(r,p)dL(u)dS(r),$$

$$F_2(p) = \int_S \left(\varepsilon(r)e_b(r)\tau(r,p) + L_\sigma(r,p)\right) K(r,p)dS(r)$$

and

$$L_\sigma(r,p) = \int_{(r,p)} (a(u) + \Phi_\sigma(r,u)) e_b(u)\tau(u,p)dL(u)$$

are continuous functions.

The net heat source $q_V(p)$ is a continuous function defined in the fundamental domain $D - S$, and unique solution of the integral equation:

$$\frac{q_V(p)}{a(p)} - \frac{1}{4}\int_S \int_{(r,p)} \Phi_\sigma(r,u)\tau(u,p)\frac{q_V(u)}{a(u)}dL(u)K_V(r,p)dS(r) = F(p)$$

with

$$
\begin{aligned}
F(p) &= F_3(p) - F_4(p), \\
F_3(p) &= 4e_b(p) - \int_S H_\varepsilon(r)\sqrt{1-\varepsilon(r)}\tau(r,p)K_V(r,p)dS(r), \\
F_4(p) &= \int (\varepsilon(r)e_b(r)\tau(r,p) + L_\sigma(r,p))K(r,p)dS(r).
\end{aligned}
$$

Existence and uniqueness of the solution H_ε, hence of the divergence $q_V(p)$ of the net radiative heat flux vector, is assumed on the basis of the developments given in Chap. 4. Here, we aim at determining a reliable numerical solution of the problem above.

To be specific, we shall consider the approximation spaces $Y_{h^m(n)}(D)$ and $X_{h(n)}(S)$ defined respectively to be:

$$
\begin{aligned}
Y_{h^m(n)}(D) &= \{L_{h^m(n)}(\psi); \quad \psi \in C^o(D)\}, \\
X_{h(n)}(S) &= \{L_{h(n)}(\varphi); \quad \varphi \in C^o(S)\},
\end{aligned}
$$

with $h(n)$ being the maximum diameter of the finite surface elements in which case n is the number of surface elements used to defined a partition of S, and $h^m(n)$ being the maximum diameter of the finite volume elements being considered in which case n is the number of volume elements used to defined a partition of the entire volume D.

We are looking for the functions $H_{\varepsilon,h(n)}$ together with the associated net outflow of radiant energy per unit volume $q_{V,h^m(n)}$ in the expansion forms:

$$H_{\varepsilon,h(n)}(p) = \sum_{i=1}^{M(n)} H_{\varepsilon,h(n)}(p_i)\varphi_i(p), \quad i=1,\cdots,M(n), \quad p_i \in S$$

$$q_{V,h^m(n)}(p) = \sum_{j=1}^{M^m(n)} q_{V,h^m(n)}(p_j)\psi_j(p), \quad j=1,\cdots,M^m(n), \quad p_j \in D$$

where the φ_i and ψ_j are basis functions of the subspaces $X_{h(n)}(S)$ and $Y_{h^m(n)}(D)$, respectively.

Since $H_{\varepsilon,h(n)}$ solves the approximated variational problem associated to the irradiance formulation governed by the first integral equation of the formulation hereabove, and the function $q_{V,h^m(n)}$ satisfies the natural approximated problem governed by the second integral equation, solving our approximated problem is also equivalent to solving the following linear system of equations:

$$B \cdot H_{\varepsilon,h(n)} \;\; = \;\; L - \frac{1}{4}A \cdot q_{V,h^m(n)} \,,$$

$$(C - \frac{1}{4}E \cdot B^{-1} \cdot A)q_{V,h^m(n)} \;\; = \;\; d - E \cdot B^{-1} \cdot L\,.$$

The unknowns in the above linear system are the vectors

$$H_{\varepsilon,h(n)} \equiv \left(H_{\varepsilon,h(n)}(p_i)\right)_{1 \le i \le M(n)} \,,$$

and

$$q_{V,h^m(n)} \equiv \left(q_{V,h^m(n)}(p_i)\right)_{1 \le i \le M^m(n)} \,.$$

Here, $B \equiv \left(B(\varphi_i, \varphi_j); \; 1 \le i,j \le M(n)\right)$ is the matrix of the linear system defined in Chap. 5, and $L \equiv \left(L(\varphi_i); \; 1 \le i \le M(n)\right)$ is the corresponding right hand vector. The vector $d \equiv (d_i; \; 1 \le i \le M^m(n))$ and the following matrices:

$$\begin{aligned}
C &\equiv (C_{ij})_{1 \le i,j \le M^m(n)} \,, \\
A &\equiv (A_{ij})_{1 \le i, \le M(n), 1 \le j \le M^m(n)} \,, \\
E &\equiv (E_{ij})_{1 \le i, \le M^m(n), 1 \le j \le M(n)} \,,
\end{aligned}$$

are defined by the following relations:

$$\begin{aligned}
d_i &= (4e_b(p_i) - F_4(p_i)) \\
E_{ij} &= \int_S \sqrt{1 - \varepsilon(r)}\tau(r,p)K_V(r,p_i)\varphi_j(r)dS(r), \\
C_{ij} &= \delta_{ij} - \frac{a(p_i)}{4}\int_S L_\Phi(r,p_i)K_V(r,p_i)dS(r), \\
A_{ij} &= \int_S \int_S \sqrt{1 - \varepsilon(p)}L_3(r,p_i)K(r,p_i)\varphi_i(p)dS(r)dS(p), \\
L_\Phi(r,p_i) &= \int_{(r,p_i)} \Phi_\sigma(r,u)\tau(u,p_i)\frac{\psi_j(u)}{a(u)}dL(u).
\end{aligned}$$

In this chapter we have extended our results to scattering media. It has been shown that it is feasible to extend the mathematical algorithm developed to account for the scattering effects. However, in solving practical problems the main difficulty stem from the lack of information about the (scattering) phase function and the distribution (size and concentration) of solid particles. In the past, however, few investigations have dealt with anisotropic scattering. Most of them are limited to simple forms of the phase function and constant scattering coefficients [139]-[143].Thus, it seems to be premature to tackle this problem before obtaining the desired insight on phase functions by performing appropriate experiments/measurements.

9.2 Radiation Heat Transfer with Conduction and/or Convection

So far we have concerned ourselves with the radiative heat transfer in the absence of interaction with other mode of energy transport. In many practical systems, however, a significant amount of heat can be transferred simultaneously by conduction and/or convection, and the combined effect of the three transfer modes must be accounted for.

9.2.1 The Energy Conservation Relation

A general energy balance on a volume element includes radiation, conduction, convection, internal heat sources, compression work, viscous dissipation and energy storage due to transients. Storage of radiant energy within an element is generally negligible, radiation pressure is also neglected with respect to the fluid pressure and hence does not contribute to the compression work. Thus, for a moving fluid, the general form of the energy conservation equation is often written to be [49]:

$$\rho c_p \left(\frac{\partial T}{\partial t} + \vec{v} \cdot \overrightarrow{grad}(T) \right) = div(k \cdot \overrightarrow{grad}(T)) - q_V - P \cdot div(\vec{v}) + \Phi_\nu + \dot{Q}'''$$

with ρ being the fluid density, c_p the heat capacity, T the internal temperature, $\vec{v}$ the velocity vector of the moving fluid, k the thermal conductivity, q_V the net radiative heat source, P the pressure, Φ_ν the heat production by viscous dissipation, and $\dot{Q}'''$ the rate of heat generated per unit volume and time within the medium (such as energy release due to chemical reactions).

Some, none, or all of these terms may be present in a given radiation problem. The net radiative heat source depends upon the temperature within the medium and at the enclosure boundary. Therefore, this temperature field must be determined through the above energy conservation equation that incorporates all three modes of heat transfer.

Neglecting transient terms in the energy conservation equation, one obtains the relation:

$$\rho c_p \vec{v} \cdot \overrightarrow{grad}(T) = div(k \cdot \overrightarrow{grad}(T)) - q_V - P \cdot div(\vec{v}) + \Phi_\nu + \dot{Q}''' \qquad (9.1)$$

For uniform thermal conductivity; that is, independent of position and temperature, we shall rewrite (9.1) in the form:

$$\frac{\rho c_p}{k} \vec{v} \cdot \overrightarrow{grad}(T) = \Delta T - \frac{1}{k}\left(q_V + P \cdot div(\vec{v}) - \Phi_\nu - \dot{Q}'''\right).$$

The term $\rho c_p/k$ is known as the medium *"thermal diffusivity"*. It combines three physical quantities: the density ρ, the heat capacity c_p, and the thermal conductivity k into a single quantity.

9.2.2   Concluding Remarks

The interaction between the three transfer mode can be simple or quite complex depending on the physical situation. The simple picture, uncoupled problems, is where the heat dissipation and heat fluxes are independent. These are then computed separately and the individual results added. In very complex situations, coupled problems, as found in combustion chambers and industrial furnaces, the desired unknown quantities, local temperatures and net heat sources within the medium, cannot be determined by adding separate solutions. In these cases the energy conservation equation must be solved simultaneously with the radiative heat transfer equation. These conditions would often required the temperature at the enclosure boundary to be specified and we shall seek to determine the temperature distribution inside the enclosure volume and the radiative heat fluxes at the enclosure boundary and volume.

10
Conclusion

Radiation is the dominant mechanism of heat transfer in high temperature industrial furnaces, combustion chambers and boilers. We have made an effort to develop a coherent and comprehensive treatise to the radiative heat transfer in enclosures. This book provides a detailed theoretical examination of the radiative heat transfer equation and dwells on a unique and the most reliable method for calculating radiative heat exchanges in enclosures.

We have seen that the radiative heat transfer problems in enclosures can be classified into two categories:

- The determination of radiant energy exchanges throughout a given medium in which the temperature distributions, the species concentration, the absorption, scattering, and emission properties of the medium are given;

- The determination of temperature distribution and radiating energy exchanges in a given volume.

Our task here was centered on the first point concerning the solution of a complex integro-differential equation.

As a consequence of its inherently intractable form, the radiation transfer equation has long been a subject of many studies which have led to a collection of many approximate solution procedures. In the course of such quests, the radiative heat transfer equation was subjected to many uncontrolled assumptions and simplifications in order to make it amenable to numerical representations and analysis. Subsequently, the theoretical developments on the computational methods for the solution of the integro-differential equation of radiative heat transfer problem have been centered on modifications of the basic structure of the intensity of incident radiation. Engineering approximations of an ad hoc nature were made at different stages in the search for the solution. These approximations obscured both theoretical developments on the radiative heat transfer equation, and the conclusions that could be drawn from their success or failure in simulating a real enclosure. In a number of applications, in the physical world, it was required of these models to produce quantitative information about the intensity and flux of incident radiation. Their intractability under such requirements soon became evident.

In this book the radiative heat transfer equation has been used in its full generality to deduce, in quantitative details, the salient features of a beam intensity of radiation in absorbing-emitting and scattering media. The work presented has exploited a full range of mathematical tools, familiar in other fields of physics, without obscuring the physical nature of radiation. We have demonstrated that all the parameters describing the radiative heat transfer can be determined exactly, knowing the necessary inputs, for any enclosure situations. We have formulated and solved a structural problem for heat transfer by radiation. We have proved the existence and uniqueness of the solution of the radiative heat transfer equation and we have shown that radiative heat transfer in enclosures is an elliptic boundary value problem. This observation may lead to new insights. In particular we believe that the study of eigenvalue problems associated with the integro-differential operator of the radiative heat transfer equation may result in further advancing the understanding of this phenomenon.

Existence of a solution shows that no mutually contradictory conditions are demanded to the solution, and uniqueness of the solution shows that the problem of radiative transfer is completed. Since the method of solution developed offers high accuracy, it can be used for benchmark assessment of radiative heat transfer calculations in enclosures of any shapes, filled up with any temperature distribution.

While designing combustion chambers, not always all the necessary parameter, such as the boundary surface emissivity, in-furnace temperature distributions, and the medium composition, are available with a good accuracy. Nevertheless, the analysis has shown that for a small variation in the input data, there corresponds a small variation in the solution of our radiation problem. Thus, the procedure developed can be used for the design and operation of industrial combustion chambers. For any generated temperature distributions within the enclosure, the procedure can be used to determined the heat fluxes at the enclosure boundary, and the heat sinks within the enclosure volume. Radiation intensities at any point of the enclosure and in every single direction can also be calculated. Using a multigrid approach, the method developed can be incorporated into other computer programs that either perform an overall energy balance of a system or simulate fluid flow and heat transfer problems.

The book contains several examples of radiative exchange calculations in emitting-absorbing media of gray gases; e.g. a single absorption coefficient of constant value was used to represent emission and absorption. We have made comparisons of method predictions against incident radiative heat flux measurements for natural gas and propane flames at an industrial scale. Thus, the usefulness of the procedure developed for computing radiative exchange in emitting-absorbing media has been demonstrated. Our considerations on scattering have indicated that it is possible to include this phenomenon into the algorithm developed. Only one example contains spectral calculations in a volume of a homogeneous gas surrounded by isothermal walls. Some practical problems, however, require procedures able to handle strongly non-homogeneous gaseous mixtures.

Notwithstanding the merits of engineering correlations available in the literature, we call attention to the need for a proper methodology for calculating the coefficient of absorption locally inside the enclosure volume. For gas filled enclosures with non-isothermal temperature and radiating species distributions, we cannot deduce an adequate formula for the absorption coefficient, which depends only on the state variables, merely from the developments given in Chap. 7. Also we did not wanted to loose ourself into speculations and discussions that arise from this search, but merely remark that knowledge of the medium absorption coefficient allows the derivation of the solution of to boundary value problem.

There is no lack of textbooks on heat transfer by radiation. The applied research scientist or the engineer in radiative heat transfer will have difficulty in finding a discussion that leads him straight to the heart of the subject and gives him the power to simulate very efficiently radiation heat transfer in enclosures. We make no claim to have presented the engineers and researchers in radiative heat transfer with the ideal textbook. Yet we do not consider the present developments superfluous. In order and choice of material, in fundamental aim, and in mode of presentation, this book differs considerably from the current literature.

As mentioned already in the introductory chapter, the writing of a treatise in radiative heat transfer is a rather difficult task for which some years of meditation are necessary. The book must satisfy two requirements which are not easily reconciled. From the mathematical point of view, it must be written in accordance with standards of mathematical rigor and precision. From the radiative heat transfer point of view, it must not be written so abstractly as to discourage physicists and engineers who need this mathematical tool. Have we succeeded in satisfying both requirements? Only the reader can judge.

Nomenclature

$1_\Omega\left(\cdot\right)$	:	Characteristic function of the set Ω	-
$a,\ a\left(\cdot\right)$	:	Medium absorption coefficient	m^{-1}
$a_n\left(\cdot,\cdot\right)$	:	Mie scattering coefficient	-
A	:	Linear operator	-
	:	Matrix of a linear system of equations	-
$A\left(\cdot,\cdot,\cdot,\cdot\right)$	:	Total band absorptance	cm^{-1}
$b\left(\cdot\right)$	:	Linear transformation	-
$b_n\left(\cdot,\cdot\right)$	:	Mie scattering coefficient	-
B	:	Matrix of a linear system	-
B^t	:	Transpose of the matrix B	-
$B\left(\cdot,\cdot\right)$	:	Bilinear form	-
$B\left(\cdot,\cdot,\cdot,\cdot\right)$	:	General function	-
c	:	Speed of light	m/s
	:	Positive constant	-
C	:	Matrix of a linear system	-
C_1, C_2, C_3	:	Radiation constants	$W\cdot m^2,\ \mu m\cdot K$
$C\left(\cdot,\cdot,\cdot,\cdot\right)$	:	General function	-
$C_{\max}$	:	Maximal curvature	-
d	:	Polynomial degree	-
	:	Order of numerical integration	-
	:	Mean spectral line spacing	μm
$d\left(\cdot,\cdot\right)$	:	Distance operator	-
D	:	Physical domain	-
$D\left(\cdot,\cdot,\cdot,\cdot\right)$	:	General function	-

$dL\left(\cdot\right)$	:	Lebesgue measure	-
	:	Differential small path length	m
$dS\left(\cdot\right)$	:	Lebesgue measure	-
	:	Differential small surface	m^2
$e_b\left(\cdot\right), e_b\left(\cdot,\cdot\right)$	:	Blackbody emissive power	-
$E_b\left(\cdot\right)$	:	Blackbody emissive power	-
E	:	Matrix of a linear system	-
	:	Radiative energy, photon energy	W
$E\left(\cdot,\cdot,\cdot\right)$	:	Radiative energy, photon energy	W
$E\left(\cdot,\cdot\right)$	:	Efficiency factor for emission	-
$F\left(\cdot\right)$	:	Continuous function	-
$F\left(\cdot,\cdot,\cdot,\cdot\right)$	:	General function	-
$G\left(\cdot\right)$	:	Quadratic functional, energy integral	-
$G\left(\cdot,\cdot\right)$	:	Source intensity of radiation	-
$G\left(\cdot,\ldots,\cdot\right)$	:	General functions	-
h	:	Planck's constant	$J \cdot s$
	:	Positive parameter	-
$h\left(\cdot\right)$	:	Positive size parameter	-
$H\left(\cdot\right)$	:	Surface irradiance function	W/m^2
$H_\varepsilon\left(\cdot\right)$	:	Surface irradiance function	W/m^2
$H_{\varepsilon,h(n)}\left(\cdot\right)$	:	Approximation of $H_\varepsilon\left(\cdot\right)$	W/m^2
$H\left(\cdot,\cdot\right)$	:	Continuous function	-
$I\left(\cdot,\cdot\right)$	:	Total intensity of radiation	$W/m^2 \cdot sr$
$I\left(\cdot,\cdot,\cdot\right)$	:	Radiation spectral intensity	$W/m^2 \cdot sr \cdot \mu m$
$\inf\left(\cdot\right)$	:	Lower bound function	-
$J\left(\cdot\right)$	:	Surface radiosity function	W/m^2
	:	Quadratic functional	-
k	:	Boltzmann constant	J/K
$k\left(\cdot\right)$	:	Contour function of a spectral line	-
$K\left(\cdot,\cdot\right)$	:	Kernel function	-
$K_o\left(\cdot,\cdot\right)$	:	Kernel function	-
$K_V\left(\cdot,\cdot\right)$	:	Kernel function	-
$l\left(\cdot\right)$	:	Linear transformation	-
	:	Length of arc	-
L	:	Vector of a linear system	-
$L\left(\cdot\right)$	:	Linear form	-
$L\left(\cdot,\cdot\right)$	:	Line integral	-
$L\left(\cdot,\cdot,\cdot\right)$	:	Spectral line integral	-
$L\left(S_k,\cdot\right)$	:	Lagrange $P\left(S_k\right)$-interpolant function	-
$L_{h(n)}\left(\cdot\right)$	:	Lagrange interpolant	-
m	:	Particle index of refraction	-
$\max\left(\cdot\right)$	:	Maximum function	-
$\min\left(\cdot\right)$	:	Minimum function	-
$mes\left(\cdot\right)$	:	Measure function	-

M	:	Photon momentum	$W \cdot s/m^2$
	:	Order of multiplicity of an eigenvalue	-
n	:	Medium refractive index	-
	:	Finite dimension of a physical space	-
$N\left(\cdot,\cdot\right)$	:	Kernel function	-
$N\left(S_k\right)$	:	Finite set of distinct points of S_k	-
p	:	Physical point of an enclosure	-
p_i	:	Nodal point, quadrature point	-
$\mathrm{proj}_{\perp(r,p)}\left(\cdot\right)$	:	Projection on to the orthogonal of (r,p)	-
P	:	Total pressure of gas	Pa
P_e	:	Pressure broadening	Pa
$P\left(\cdot\right)$	:	Radiation pressure function	-
$q\left(\cdot\right)$	:	Net radiative heat flux function	W/m^2
$q_{h(n)}\left(\cdot\right)$	:	Approximation of net heat flux	W/m^2
$q_V\left(\cdot\right)$	:	Net outflow of radiant energy	W/m^3
Q	:	Total heat flow	W
$Q_m\left(\cdot\right)$	:	Quadrature operator	-
r	:	Physical point of an enclosure	-
r_i	:	Nodal point, quadrature point	-
(r,p)	:	Line of sight from point r to point p	-
$\overrightarrow{rp}$	:	Direction of popagation of radiation	-
R	:	Reference triangular element	-
$\Re\left(\cdot\right)$	:	Univocal relation	-
$R\left(S_k,\cdot\right)$	:	Restriction operator onto S_k	-
s	:	Physical point of an enclosure	-
S	:	Enclosure boundary (inner surface)	-
S_o	:	Parameter of an exponential distribution	-
$Sup\left(\cdot\right)$	:	Supremum function	-
$Supp\left(\cdot\right)$	:	Support function	-
$S\left(\cdot,\cdot\right)$	:	Efficiency factor for scattering	-
$S\left(\cdot,\cdot,\cdot\right)$	:	Strength of a particular spectral line	-
t	:	Time variable	s
T	:	Enclosure surface temperature	K
T_o	:	Reference temperature	K
$T\left(\cdot\right)$	:	Enclosure temperature distribution	K
u	:	Physical point of an enclosure	-
$u\left(\cdot,\cdot,\cdot\right)$	:	General function	-
$U\left(\cdot,\cdot,\cdot,\cdot\right)$	:	General function	-
V	:	Enclosure volume	m^3
x	:	Cartesian coordinate	m
	:	Particle size parameter	μm
$X\left(\cdot,\cdot\right)$	:	Mass path length	g/m^2
$\bar{X}\left(\cdot,\cdot\right)$	:	Scaled mass path length	g/m^2
y	:	Cartesian coordinate	m
z	:	Cartesian coordinate	m

Greek charcters

α	:	Enclosure surface absorptivity	-
$\alpha\left(\cdot\right)$	:	Constant of coercivity	-
$\alpha\left(\cdot\right)$	:	Integrated band intensity	$cm^{-1}/g\cdot m^{-2}$
$\bar{\alpha}\left(\cdot\right)$	:	Scaled integrated band intensity	$cm^{-1}/g\cdot m^{-2}$
β	:	Line-width to spacing parameter	-
$\beta\left(\cdot\right)$	:	Extinction coefficient	m^{-1}
$\beta\left(\cdot,\cdot\right)$	:	Pressure broadening parameter	-
$\bar{\beta}\left(\cdot,\cdot\right)$	:	Scaled pressure broadening parameter	-
$\chi\left(\cdot,\cdot\right)$	:	Indicative function of shadow zones	-
$\delta\lambda,\delta\eta$	:	Narrow bands	-
	:	width of narrow bands	$\mu m,\ cm^{-1}$
δ_{ij}	:	Kronecker symbol	-
ν	:	Frequency of a photon energy	Hz
$\left|\delta\lambda\right|,\left|\delta\eta\right|$	:	width of spectral intervals	$\mu m,\ cm^{-1}$
ε	:	Constant enclosure surface emissivity	-
	:	Arbitrary positive constant	-
$\varepsilon\left(\cdot\right)$	:	Enclosure surface emissivity function	-
ϕ,φ,Φ	:	General functions or variables	-
γ,γ_i	:	Scalars	-
η	:	Arbitrary positive number	-
	:	Wavenumber	cm^{-1}
$\kappa\left(\cdot,\cdot\right)$	:	Optical depth of the medium	-
λ	:	Wavelength	μm
μ	:	Eigenvalue	-
$\theta\left(\cdot,\cdot,\cdot\right)$	:	Scattered angle	-
$\rho\left(\cdot\right)$	:	Density of a gas species	kg/m^3
	:	Diameter function	m
σ	:	Constant scattering coefficient	m^{-1}
	:	Stefan-Boltzmann constant	$W/\left(m^2K^4\right)$
$\sigma\left(\cdot\right)$	:	Scattering coefficient function	m^{-1}
$\tau\left(\cdot,\cdot\right)$	:	Transmissivity function	-
$\tau\left(\cdot,\cdot,\cdot\right)$	:	Spectral transmissivity function	-
ω_i	:	Coefficient of quadrature formula	-
ω_o	:	Band width parameter	cm^{-1}
$\omega\left(\cdot\right)$	:	Band width parameter	cm^{-1}
$\bar{\omega}\left(\cdot\right)$	:	Scaled band width parameter	cm^{-1}
$\omega\left(\cdot,\cdot\right)$	:	Solid angle	$sr.$
$\Omega\left(\cdot\right)$	:	Albedo for scattering	-
ξ,ζ	:	General variables or functions	-

Functional and physical spaces

$H^2(S)$	:	Second order Sobolev space -
$(\cdot,\cdot)_{H^2(S)}$	:	Inner product defined on the space $H^2(S)$ -
$\|\cdot\|_{H^2(S)}$	:	Norm induced by the inner product $(\cdot,\cdot)_{H^2(S)}$ -
$\|\cdot\|_{H^2(S)}$	:	Semi norm defined on the space $H^2(S)$ -
$L^2(S)$	:	Set of measurable real-valued and square
	:	integrable functions acting on S -
$(\cdot,\cdot)_{L^2(S)}$	:	Inner product defined on the space $L^2(S)$ -
$\|\cdot\|_{L^2(S)}$	:	Norm induced by the inner product $(\cdot,\cdot)_{L^2(S)}$ -
$P(S_k)$	:	Vectorial space of real-valued functions on S_k -
$\mathbb{R}^3$	:	Physical space -
$\|\cdot\|_{\mathbb{R}^3}$	:	Euclidean distance defined on $\mathbb{R}^3$ -
$\|x\|$	:	Module of the quantity x -
$\Omega_{h(n)}(S)$	:	Regular set of triangulation of S -
$|\Omega|$	:	Cardinal of the set Ω -
$\mathbb{R}^n$	:	Space of n tuples -
$X_{h(n)}(S)$	:	Approximated space of finite dimension -
$\left|X_{h(n)}(S)\right|$	:	Finite dimension of the space $X_{h(n)}(S)$ -

Chemical species

CO	:	Carbon monoxide -
CO_2	:	Carbon dioxyde -
CH_4	:	Methane -
C_xH_y	:	Hydrocarbon -
H_2	:	Hydrogen -
H_2O	:	Water -
N_2	:	Nitrogen -
NO	:	Nitric oxide -
NO_x	:	Oxides of nitrogen -
O_2	:	Oxygen -

Superscripts - Subscripts

a	:	Absorbed
b	:	Blackbody
e	:	Emitted
i	:	Incoming, incident
m	:	Medium
o	:	Outgoing (emitted plus reflected)
t	:	Transpose

References

[1] Mbiock A. (1997) Radiative Heat Transfer in Furnaces - Elliptic Boundary Value Problem. PhD Thesis in Applied Mathematics, Rouen University, France

[2] Siegel R., Howell J. R (1981) Thermal Radiation Heat Transfer. 2nd edn. Hemisphere, Washington D.C.

[3] Hottel H. C. (1931) Radiant Heat Transmission Between Surfaces Separated by Non-absorbing Media. Trans. ASME, 53: 265-273

[4] Hottel H. C. (1933) Effect of Reradiation on Heat Transmission in Furnaces and Through Openings. Trans. ASME, 55: 39-49

[5] Thring M. W. (1948) Proposals for the Establishment of an International Research Project on Luminous Radiation. The British Iron and Steel Research Association Physics Depatment, Laboratory: 140 Battersea Park Road, London, S. W. 11., Physics Note No. 40.

[6] De Graaf J. E. (1975) The History of International Flame Research. IFRF, The Netherlands

[7] Thring M. W. (1989) Past, Present and Future of the IFRF. In: proceedings of the IFRF 9th Members' Conference. IFRF, The Netherlands

[8] Weber R. (1998) The Spirit of IJmuiden. Fifty Years of the IFRF. IFRF, IJmuiden, The Netherlands

[9] Selçuk N. (1985) Exact Solutions for Radiative Heat Transfer in Box-shaped Furnaces. Journal of Heat Transfer, Trans. ASME, 107(3): 648-655

[10] Selçuk N., Tahiroglu Z. (1988) Exact Numerical Solutions for Radiative Heat Transfer in Cylindrical Furnaces. Int. Journal for Numerical Methods in Engineering, 26: 1201-1212

[11] Gritton E. C., Leonard A. (1970) Exact Solutions to the Radiation Heat Transport Equation in Gaseous Media Using Singular Integral Equation Theory. Journal of Quant. Spectrosc. and Radiative Transfer, 10: 1095-1118

[12] Howell J. R. (1988) Thermal Radiation in Participating Media: the past, the present and some futures. Journal of Heat Transfer, Trans. ASME, 110(4):1220-1229, 1988.

[13] Tong T. W., Skocypec R. D. (1992) Summary on Comparison of Radiative Heat Transfer Solutions for a Specified Problem. ASME J. Heat Transfer, 203:253-264

[14] Sparrow E. M. (1960) Application of Variational Methods to Radiation Heat Transfer calculations. ASME J. Heat Transfer, 82c:375-380

[15] Sparrow E. M., Haji-Sheikh (1965) A Generalized Variational Method for Calculating Radiant Interchange Between Surfaces. ASME J. Heat Transfer, 82c: 103-109

[16] Breitbach G., Altes J., Eczmarowsky M. (1990) Solution of Radiative Problems using Variational based Finite Elements Method. International Journal for Numerical Methods in Engineering, 29: 1701-1714

[17] Courant R., Hilbert D. (1989) Method of Mathematical Physics, Volumes 1 and 2. Wiley Classic Edition, Interscience publishers, New York

[18] Raviart P. A., Thomas J. M. (1992) Introduction à l'Analyse Numérique des Equations aux Dérives Partielles. Masson, Paris

[19] Dautray R., Lions J. L. (1984) Analyse Mathmatique et Calcul Numérique, Volumes 1 to 9. Collection Enseignement, INSTN CEA, Paris

[20] Courant R. (1964) Differential and Integral Calculus. Blackie and Son Limited, London

[21] Mikhlin S. G., Smolitsky K. L. (1967) Approximate Methods for Solution of Differential and Integral Equations. American Elsevier, New-York

[22] Tricomi F. G. (1965) Integral Equations. Interscience Publishers, Inc., New York

[23] Hochstadt H. (1973) Integral Equations. Wiley Classic Edition, Interscience Publishers, New York

[24] Yosida K. (1960) Lectures on Differential and Integral Equations. Interscience Publishers, Inc., New York

[25] Pogorzelski W. (1966) Integral Equations and their Applications. Pergamon Press, Oxford

[26] Mikhlin S. G. (1965) Integral Equations. Series in pure and applied mathematics, Pergamon Press, Oxford.

[27] Halmos P. R. (1950) Measure Theory. The University Series in Higher Mathematics, D Van Nostrand Company, Inc., New York

[28] Hildebrant T. H (1963) Introduction to the Theory of Integration. Academic Press Inc., London

[29] Jean R. V. (1989) Mesure et Integration. Presses de l'Université du Qubec

[30] Guichardet A. (1989) Integration Analyse Hilbertienne. X Ecole Polytechnique, Edition Marketing, Paris.

[31] Sobolev S. L (1936) Méthode Nouvelle à Résoudre le Problème de Cauchy pour les Equations Lineaires Hyperboliques Normales. Mat. Sbornik, volume 1, Paris

[32] Schwartz L. (1950) Thorie des Distributions, volume 1. Hermann, Paris

[33] Laurent P. J (1972) Approximation et Optimization. Collection Enseignement des Sciences, 13, Hermann, Paris

[34] Mikhlin S. G. (1970) Mathematical Physics, an Advanced Course. North-Holland Publishing Company, Amsterdam

[35] Ciarlet P. G. (1994) Introduction à l'Analyse Numérique Matricielle et à l'Optimization. Masson, Paris

[36] Mikhlin S. G. (1969) Variational Methods in Mathematical Physics. Series in pure and applied mathematics, Pergamon Press, Oxford.

[37] Mikhlin S. G. (1971) The Numerical Performance of Variational Methods. Noorhoff, Groningen, The Netherlands

[38] Brezis H. (1992) Analyse Fonctionnelle. Masson, Paris

[39] Dunford N., Schwartz J. T. (1964) Linear Operators, Part two, Spectral Theory-Self Adjoint Operators in Hilbert Space. Interscience Publishers, Inc., New York.

[40] Brebbia C. A., Wendland W. L., Kuhn G. (1987) Boundary Element IX: Vol 1: Mathematical and Computational Aspects. Computational Mechanics Publications, Springer-Verlag

[41] Brebbia C. A., Zamani N. G. (1989) Boundary Element Techniques: Theory and Applications in Engineering. Computational Mechanics Publications, Southampton, Boston

[42] Brebbia C. A., Dominguez J. (1989) Boundary Elements - An Introductory Course. Computational Mechanics Publications, McGraw-Hill Book Co.

[43] Aliabadi M. H., Brebbia C. A. (1993) Advanced Formulations in Boundary Elements Methods. Computational Mechanics Publications, Elsevier Applied Science

[44] Banerjee P. K. (1994) Boundary Elements Methods in Engineering. McGraw-Hill Book Co., London Science

[45] Bialecki R. A. (1993) Solving Heat Radiation Problems Using the Boundary Element Method. Topic in Engineering, Vol. 15, Computational Mechanics Publications

[46] Van de Hulst H. C. (1981) Light Scattering by Small Particles. Dover Publications, New York

[47] Abramowitz M., Stegun I. A. (1965) Handbook of Mathematical Functions. Dover publications, New-York

[48] Lowes T. M., Heap M. P. (1971) Emission-Attenuation Coefficients of Luminous Radiation. International Flame Research Foundation, Proceedings of the IFRF 2nd members' Conference. IFRF, The Netherlands

[49] Modest M. F. (1993) Radiative Heat Transfer. McGraw-Hill, Inc., New York

[50] Schuster A. (1905) Radiation through a Foggy Atmosphere, Astrophysics J., 21(4):1-22

[51] Siddall R. G. (1972) Flux Methods for the Analysis of Radiant Heat Transfer. Fourth Symposium on Flames and Industry, British Flame Research Committee, 1972. Published also in the Journal the Institute of Fuel, 2:169-177

[52] Patankar S. V., Spalding D. B. (1973) A Computer Model for Three-Dimensional Flow in Furnaces. Fourteenth International Symposium on Combustion. The Combustion Institute, 605-614

[53] Gossman A. D., Lockwood F. C. (1973) Incorporation of a Flux Model for Radiation into a Finite-Difference Procedure for Furnace Calculations. Fourteenth International Symposium on Combustion. The Combustion Institute, 661-671

[54] Lowes T. M., Bartelds H., Heap M. P., Michelfelder S., Pai B. R. (1973) Prediction of Radiant Heat Transfer in Axisymmetrical Systems. IFRF Doc. G02/a/25. IFRF, The Netherlands

[55] Bartelds H., Heap M. P., Lowes T. M. (1977) Radiative Heat Transfer in Enclosures. IFRF Doc. G04/a/6, IFRF, The Netherlands

[56] Fiveland W. A. (1984) Discrete Ordinates Solution of the Radiative Transport Equation for Rectangular Enclosures. J. Heat Transfer. Trans. ASME, 106(4)669-706

[57] Truelove J. S. (1987) Discrete Ordinates Solutions of the Radiation Transport Equation. J. Heat Transfer. Trans. ASME, 109(4):1048-1051

[58] Truelove J. S. (1988) Three-Dimensional Radiation in Absorbing-Emitting-Scattering Media using the Discrete Ordinates Approximation. J. Quant. Spect. Radiat. Heat Transf., 39(1):27-31

[59] De Marco A. G., Lockwood F. C (1975) A New Flux Model for the Calculation of Radiation in Furnaces. Italian Flame Day, La Rivisita dei Combustibili, 29:184-196

[60] Lockwood F. C., Shah G. (1978) Evaluation of an Efficient Radiation Flux Model for Furnace Prediction Procedures. Heat Transfer, 2:33-38

[61] Carlson B. G., Lathrop K. D. (1968) Transport Theory: The Method of Discrete Ordinates. In: Computing Methods in Reactor Physics, Gordon and Breach Science Publishers, New-York.

[62] Gelbard E. M. (1968) Spherical Harmonics Methods: PL and Double-PL Approximations. In: Computing Methods in Reactor Physics, Gordon and Breach Science Publishers, New-York.

[63] Fengshan L., Swithenbank J., Garbett E. S. (1992) The Boundary Condition of the PN-Approximation used to Solve the Radiative Transfer Equation. Int. Journal of Heat and Mass Transfer, 35(8)2043-2052

[64] Lockwood F. C., Shah G. (1981) A New Radiation Solution Method for Incorporation in General Combustion Procedues. Eighteenth Int. Symp. on Comb., The Combustion Institute, 18:1405-1414

[65] Doherty P., Fairweather M. (1988) Predictions of Radiative Transfer from Nonhomogeneous Combustion Products Using the Discrete Transfer Method. Combustion and Flame, 71:79-87

[66] Carvalho M. G., Olivera P., Semiao V. (1988) A Three-Dimensional Modelling of an Industrial Glass Furnace. Journal of the Institute of Energy, 12:143-156

[67] Carvalho M. G., Coelho P. J. (1989) Heat Transfer in Gas Turbine Combustor. Journal of Thermophysics and Heat Transfer, 2(2):123-131

[68] Brewster M. Q. (1992) Thermal Radiative Transfer and Properties. A Wiley-Interscience publication, John Wiley and Sons, Inc., New York

[69] Viskanta R., Mengüc M. P. (1987) Radiation Heat Transfer in Combustion Systems. Prog. Energy and Comb. Sciences, 13:97-160

[70] Carvalho M. G., Farias T. (1998) Heat Transfer in Radiating and Combusting Systems. Trans. IChemE, 76:175-184

[71] Bialecki R. A., Weber R. (1991) Heat Transfer in Industrial Furnaces. IFRF Doc G00/y/4. IFRF, The Netherlands

[72] Poljak G. (1935) Analysis of the Heat Exchange by Radiation Between Gray Surfaces by the Saldo-Method. Tech. Physics URSS, 1(5-6):555-590

[73] Shreider (1967) The Monte Carlo Method. Series in pure and applied mathematics, Pergamon press, Oxford.

[74] Vercammen A. J., Fromment G. F. (1980) An Improved Zone Method using Monte Carlo Technique for the Simulation of Radiation in Industrial Furnaces. International Journal of Heat and Mass Transfer, 23(3):329-337

[75] Audic S., Frisch H. (1993) Monte Carlo Simulation of Radiative Transfer Problem in a Random Medium: Application to a Binary Mixture. J. Quant. Spectrosc. Radiat. Transfer, 50(2):127-147

[76] Maltby J. D., Burnd P. J. (1991) Performance, Accuracy and Convergence in a Three-Dimensional Monte Carlo Radiative Heat Transfer Simulation. Numerical Heat Transfer, Part B, 19:191-201

[77] Johnson T. R., Lowes T. M., Beér J. M. (1974) Comparison of Calculated Temperatures and Heat Flux Distributions with Measurements in the IJmuiden Furnace. J. Inst. Fuel 47:39-51

[78] Hottel H. C., Sarofim A. F. (1967) Radiative Transfer. McGraw-Hill Company, New York

[79] Bialecki R. A. (1985) Applying Boundary Element Method to Calculations of Temperature Field in Bodies Containing Radiating Enclosures. In: Brebbia C. A., Maier G. (eds.) Boundary Elements VII, Springer-Verlag, Berlin, 2:35-50

[80] Bialecki R. A. (1988) Heat Transfer in Cavities: Boundary Element Method Solution. In: C A Brebbia C. A., Maier G. (eds) Boundary Elements X, Springer-Verlag, Berlin, 2:246-256

[81] 70. Bialecki R. A. (1989) Modeling 3D Band Thermal Radiation in Cavities using Boundary Elements Method. In: Brebbia C. A., Connor J. J. (eds.) Advances in Boundary Elements, Field and Flow Solutions. Springer-Verlag, Berlin, 2:116-135

[82] Bialecki R. A. (1990) Solving 3D Heat Radiation Problems in Cavities Filled by a Participating Non-gray Medium using Boundary Element Method. In: Wrobel L. C., Brebbia C. A. (eds) Computational Methods in Heat Transfer - First International Conference on Innovative Numerical Techniques in Heat Transfer. Springer-Verlag, Berlin and New York, 2:205-225

[83] Bialecki R. A. (1991) Applying the Boundary Element Method to the Solution of Heat Radiation Problems in Cavities Filled by a Non-Gray Emitting-Absorbing Medium. Numerical Heat Transfer, part A, 20:41-64

[84] Bialecki R. A. (1992) Solving Nonlinear Heat Transfer Problem using the Boundary Element Method. In: Wrobel L. C., Brebbia C. A. (eds) Boundary Element Methods in Heat Transfer - International series in Computational Engineering. Elsevier Applied Science, London, 87-122

[85] Bialecki R. A. (1992) Boundary Element Calculations of the Radiative Heat Sources. In: Wrobel L. C., Brebbia C. A., Nowak A. J. (eds) Advanced Computational Methods in Heat Transfer II - Second International Conference on Innovative Numerical Techniques in Heat Transfer. Elsevier Applied Science, London, 1:205-217

[86] Nowak A. J. (1992) Solving Coupled Problems Involving Conduction, Convection and Thermal Radiation. In: Wrobel L. C., Brebbia C. A., Nowak A. J. (eds) Advanced Computational Methods in Heat

Transfer II - Second International Conference on Innovative Numerical Techniques in Heat Transfer. Elsevier Applied Science, London, 1:145-173

[87] Razzaque M. M, Klein D. E, Howell J. R. (1983) Finite Element Solution of Radiative Heat Transfer in a Two-Dimensional Rectangular Enclosure with Gray Participating Media. Journal of Heat Transfer, Trans. of ASME, 105(4):933-936

[88] Lallemant N., Weber R. (1993) Radiative Property Models for Computing Non-Sooty Natural Gas Flames. IFRF Doc. G08/y/2. IFRF, The Netherlands

[89] Lallemant N., Sayre A., Weber R. (1996) Evaluation of Emissivity Correlations for H_2O-CO_2-N_2/Air Mixtures and Coupling with Solution Methods of the Radiative Equation. Prog. Energy Comb. Science, 22:543-574

[90] Tiihonen T. (1996) Stefan-Boltzmann Radiation on Non-Convex Surface. Laboratory of Scientific Computing, University of Jyväskylä, Finland

[91] Qatanani N. (1996) Lösungsverfahren und Analysis der Integralgleichung für das Holraum-Strahlungs-Problem. Ph.D thesis, Stuttgart University

[92] Schwab C., Wendland W. L. (1992) Kernel Properties and Representations of Boundary Integral Operators. Math. Nach. 156:187-218

[93] Wendland W. L. (1990) Boundary Elements Methods for Elliptic Problems. In: Schatz A. H., Thome V., Wendland W. L. (eds) Mathematical Theory of Finite and Boundary Element Methods. Birkhäuser Verlag, Berling.

[94] Hackbusch W. (1985) Multi-Grid Methods and Applications. Springer Verlag, New-York

[95] Davis P. J., Rabinowitz P. (1983) Methods of Numerical Integration, 2nd edn. Computer Science and Applied Mathematics, Academic Press, Inc., New-York

[96] Banerjee P. K., Butterfield R. (1979) Developments in Boundary Element Methods - 1. Elsevier Applied Science Plubishers, London

[97] Banerjee P. K., Shaw R. P. (1982) Developments in Boundary Element Methods - 2. Elsevier Applied Science Plubishers, London

[98] Banerjee P. K., Mukherjee S. (1984) Developments in Boundary Element Methods - 3. Elsevier Applied Science Publishers, London

[99] Schwab C., Wendland W. L. (1985) 3D BEM Numerical Integration. In: Brebbia C. A., Maier (eds) Boundary Elements VII. Proceedings of the 7th International Conference, Springer-Verlag, Berlin, 2:85-101

[100] Wendland W. L. (1987) Strongly Elliptic Boundary Integral Equations. State of the Art in Numerical Analysis. In: Proceedings of the Joint IMA/SIAM Conference, Oxford University Press, 511-561

[101] Wendland W. L., Yu De-Hao (1988) Adaptive Boundary Element Methods for Strongly Elliptic Integral Equations. Numerische Mathematik, 53(5):539-558

[102] Wendland W. L. (1991) Analytical and Numerical Developments in 3D Boundary Element Methods for Elastic Problems. Computer Methods in Applied Mechanics and Engineering, 91(1-3):1229-1235

[103] Wendland W. L. (1983) Boundary Element Methods and their Asymptotic Convergence. In Filippi P. (ed) Theoretical Acoustics and Numerical Techniques, CISM Courses and lectures No. 277, Springer-Verlag, New York,135-216

[104] Arnold D. N., Wendland W. L. (1983) On the Asymptotic Convergence of Collocation Methods. Journal of Computational Mathematics, 41:349-381

[105] Arnold D. N., Wendland W. L. (1985) The Convergence of Spline Collocation for Strongly Elliptic Equations on Curves. Numerische Mathematik, 47(3):317-341

[106] Wendland W. L. and Yu De-Hao (1992) A-Posteriori Local Error Estimates of Boundary Element Methods with some Pseudo-differential Equations on Closed Curves. Journal of Computational Mathematics, 10(3):273-289

[107] Bernardi C., Maday Y. (1992) Approximation Spectrales de Problmes aux Limites Elliptiques. Mathmatiques et Application 10, Springer-Verlag, Paris

[108] Zienkiewicz O. C., Taylor R. L. (1989) The Finite Element Method, Fourth Edition Volume 1, Basic Formulation and Linear Problems. Mc Graw-Hill Book Company, New York

[109] Zienkiewicz O. C., Morgan K. (1983) Finite Elements and Approximation. John Wiley, New York

[110] Dhatt G., Touzot G. (1984) Une Présentation de la Méthode des Eléments Finis, 2nd edn. Colection Universit de Compigne. Maloine S.A. Editeur, Paris

[111] Oden J. T., Reddy J. N. (1976) An Introduction to the Mathematical Theory of Finite Elements. Pure and applied mathematics, A Wiley-Interscience publication, New York

[112] Prener P. M. (1975) Splines and Variational Methods. Wiley Classic Edition, Interscience Publishers., New York

[113] Schatz A. H. (1990) An Analysis of the Finite Element Methods for Second Order Elliptic Boundary Value Problems. An Introduction. In : Schatz A. H., Thome V., Wendland W. L (eds) Mathematical Theory of Finite and Boundary Element Methods. BirkhŠuser Verlag, Berling

[114] Stroud A. H. (1971) Approximate Calculation of Multiple Integrals. Library of Congress Catalog Card No. 77-159121. Prentice-Hall, Inc., New York

[115] Sparrow E. M., Albers L. U., Eckert E. R. G. (1962) Thermal Radiation Characteristics of Cylindrical Enclosures. ASME Journal of Heat Transfer, 84C:73-81

[116] Alfano G. (1972) Apparent Thermal Emittance of Cylindrical Enclosures with and without Diaphragms. Int. Journal of Heat and Mass Transfer, 15(12):2671-2674

[117] Alfano G., Sarno A. (1975) Normal and Hemispherical Thermal Emittances of Cylindrical Cavities. ASME Journal of Heat Transfer, vol. 97, No. 3, pp. 387-390, 1975.

[118] Edwards D. K (1976) Molecular Gas Band Radiation. In: Advances Heat Mass Transfer, Academic Press, New York, 12:115-193

[119] Modak A. (1979) Radiation from Products of Combustion. Fire Research, 1:339-361

[120] Edwards D. K., Morizumi S. J. (1970) Scaling of Vibration-Rotation Band Parameters For Nonhomogeneous Gas Radiation. Journal of Quant. Spectrosc. and Radiat. Transfer, 10:175-188

[121] Felske J. D., Tien C. L. (1974) A Theoretical Closed Form Expression for the Total Band Absorptance of Infrared-Radiating Gases. Int. Journal of Heat and Mass Transfer, 17:155-158

[122] Chan S. H., Tien C. L. (1969) Total Band Absorptance of Non-Isothermal Infrared-Radiating Gases. Journal of Quant. Spectrosc. and Radiat. Transfer, 9:1261-1271

[123] Ludwig C. B., Malkmus W., Reardon J. E., Thomson J. A. L. (1973) Handbook of Infrared Radiation from Combustion Gases. National Aeronautics and Space Administration Washington, DC

[124] Penner S. S. (1959) Quantitative Molecular Spectroscopy and Gas Emissivities. Addison-Wesley publisher, New York

[125] Goody R. M. (1964) Atmospheric Radiation, Theoretical Basis, volume 1. Clarendon Press, Oxford, London E.C.4

[126] Edwards D. K., Nelson K. E. (1962) Rapid Calculation of Radiant Energy Transfer Between Nongray Walls and Isothermal H_2 and CO_2 Gas. Journal of Heat Transfer, 84(4):273-278

[127] Douglas S. H. (1965) Choice of an Appropriate Mean Absorption Coefficient for Use in the General Grey Gas Equations. Journal of Quant. Spectrosc. and Radiative Transfer, 5(1):211-225

[128] Abu-Romia M. M., Tien C. L. (1967) Appropriate Mean Absorption Coefficients for Infrared Radiation in Gases. Journal of Heat Transfer, 89(4):321-327

[129] Cess R. D., Wang L. S. (1970) A Band Absorptance Formulation For Nonisothermal Gaseous Radiation. International Journal of Heat Mass Transfer, 13:547-555

[130] Cess R. D., Tiwari S. N. (1972) Infrared Radiative Energy Transfer in Gases. In: Irvine T. F., Harnett J. P. (eds) Advances in Heat Transfer. Academic press, New York, 8:229-283

[131] Hsieh T. C., Greif R. (1972) Theoretical Determination of Absorption Coefficient and the Total Band Absorptance Including a Specific Application to Carbon Monoxide. Int. Journal of Heat and Mass Transfer, 15:1477-1487

[132] Felske J. D., Tien C. L. (1974) Infrared Radiation From Non-Homogeneous Gas Mixtures Having Overlapping Bands. Journal of Quantitative Spectroscopy in Radiative Transfer, 14:35-48

[133] Lin J. C., Greif R. (1974) Total Band Absorptance of Carbon Dioxide and Water Vapor Including Effects of Overlapping. Int. Journal of Heat and Mass Transfer, 17:793-795

[134] Lallemant N., Weber R. (1996) A Computationally Efficient Procedure for Calculating Gas Radiative Properties Using the Exponential Wide Band Model. Int. Journal of Heat and Mass Transfer, 39(15):3273-3286

[135] Michelfelder S., Lowes T. M. (1973) Report on M-2 Trials. IFRF Doc. F36/a/04, IFRF, The Netherlands.

[136] Johnson T. R. (1971) Application of the Zone Method of Analysis to the Calculation of Heat Transfer from Flames. Ph.D. thesis, Sheffield University

[137] Beér J. M., Claus J. (1962) The Traversing Method of Radiation Measurements in Luminous Flames. J. Inst. Fuel, 42:437-443

[138] Sayre A., Lallement N., Dugué J., Weber R. (1994) Effect of Radiation on Nitrogen Oxide Emissions from Non-sooty Swirling Flames of Natural Gas. The twenty-fifth symp. (international) on combustion, The Combustion Institute, 235-242

[139] Crosbie A. L., Schrenker R. G. (1982) Exact Expressions for Radiative Transfer in a Three-Dimenional Rectangular Geometry. Journal Quant. Spectrosc. Radiative Transfer, 28(6):507-526

[140] Crosbie A. L., Schrenker R. G. (1984) Radiative Transfer in a Two-Dimenional Rectangular Medium Exposed to Diffuse Radiation. Journal Quant. Spectrosc. Radiative Transfer, 31(4):339-372

[141] Cartigny J. D., Yamada Y., Tien C. L. (1986) Radiative Transfer with Dependent Scattering by Particles. ASME Journal of Heat Transfer, 108(3):608-613

[142] Tan Z. (1989) Radiative Heat Transfer in Multidimensional Emitting, Absorbing, and Anisotropic Scattering Media - Mathematical Formulation and Numerical Method. ASME Journal of Heat Transfer, 111:141-147

[143] Fengshan L., Garbett E. S., Swithenbank J. (1992) Effects of Anisotropic Scattering on Radiative Heat Transfer Using the P1-Approximation. Int. Journal of Heat and Mass Transfer, 35(10):2491-2499

Index

Printing: Saladruck, Berlin
Binding: Buchbinderei Lüderitz & Bauer, Berlin